MARS ROVER
CURIOSITY

MARS ROVER
CURIOSITY

An Inside Account
from Curiosity's Chief Engineer

ROB MANNING and WILLIAM L. SIMON

SMITHSONIAN BOOKS
WASHINGTON, DC

Published by Smithsonian Books
Director: Carolyn Gleason
Production Editor: Christina Wiginton
Edited by Gregory McNamee
Designed by Brian Barth

Library of Congress Control Number: 2014941274

ISBN 978-1-58834-473-1 (hard cover)

Manufactured in the United States of America

18 17 16 15 14   5 4 3 2 1

For Dominique and Caline
and for Victoria, Charlotte, and David,
Sheldon, Elena, and Vincent,
and in memory of Arynne

CONTENTS

Prologue   Seven Minutes of Terror   xi

Chapter 1.   The Red Planet Fixation   1

Chapter 2.   Joining the Mars Adventure   7

Chapter 3.   Avoiding Splat   19

Chapter 4.   The Birth of MSL   35

Chapter 5.   Looking Ahead to Putting People on Mars   45

Chapter 6.   The Challenges of Landing   55

Chapter 7.   The Right Kind of Crazy   65

Chapter 8.   Scientists at Work   73

Chapter 9.   The Challenges of Instrument Creation   81

Chapter 10.   Where the Devil Lives   89

Chapter 11. It's All About the People 105

Chapter 12. Bad to Worse 115

Chapter 13. Up Against a Brick Wall 129

Chapter 14. Shutdown and Restart 139

Chapter 15. The Final Stretch 153

Chapter 16. Gremlins 169

Chapter 17. On Mars! 181

Chapter 18. The Scientific Findings 197

Appendix NASA's Mars Missions 214
Acknowledgments 216
Index 218

Two planets, near twins, created simultaneously out of the same brew at the birth of the universe, both captured in orbit around the same bright star.

Yet nature has not treated them alike. One is home to a race of intelligent beings, curious and questioning, willing to spend precious resources to seek answers to questions about the origins of humankind and the origins of life.

This is a story of one giant leap in the search for answers.

FIGURE 1. An artist's depiction of the Curiosity rover, with her arm positioned for drilling. (Courtesy of NASA and JPL/Caltech)

PROLOGUE
Seven Minutes of Terror

Sunday, August 5, 2012, 10:17 p.m. I'm sitting in the crowded, chilly Cruise Mission Support Area that serves as our control room on the campus of the Jet Propulsion Laboratory in Pasadena, a short drive from downtown Los Angeles. The team members are amped up and ready. I am in a state of quiet yet buoyant Zen-like anticipation. I know this feeling well. I have been here before.

The spacecraft we are monitoring, the Mars Science Laboratory (MSL) is one of the most incredibly intricate complex of machinery, electronics, and miniaturized science labs ever created by humankind. We have spent sixty- and eighty-hour weeks dreaming up, creating, building, and testing it through the previous ten years. The product of all that effort is now approaching Mars.

In this work, it's not a question just of being good at what we do, but of being good at picking ourselves up. We live with the knowledge that failure is ingrained in the effort. But so is overcoming. What I offer in these pages is a celebration of the project and its place in humankind's quest for knowledge, and, as well, an ode to problem solving.

If everything works as designed, MSL should now be starting to perform a series of contorted maneuvers that in just seven minutes will slow it from 12,600 miles a hour (5.6 km/s) to a gentle landing on the Martian surface in a relatively smooth area that has been named Gale Crater.

These seven minutes are what we call "the seven minutes of terror."

If the landing is successful, a coded message, a string of ones and zeros, will be beamed back to us through space to confirm that all is well. It will take 13.8 minutes for that signal to reach this room.

It's as if the person you love most dearly is in a building that has been devastated by some errant force of nature and you are waiting to hear if she or he is still alive and uninjured. That's something of what it's like for those 13.8 minutes. The only hope most team members have of managing their anxiety comes from memory flashes of the vivid rehearsals we've been putting them through, rehearsals that make tonight seem more like a simulation than anything real. But we won't know for 7 + 13.8 minutes whether the lander has arrived safely and undamaged or has been crippled, smashed, or irretrievably destroyed on landing.

The control room is packed with team members anxiously huddled at their displays, watching for any sign that something is amiss. Around the world, people are crowded around television sets as NASA runs animated video of what is happening—or at least what we hope is happening. Times Square in New York is packed with people staring at the giant Jumbotron, where images of the action are on view.

It doesn't help that on these missions to Mars, we send our payloads skyward while still holding in our hands long to-do lists of essential efforts not yet tackled. Odd as it must sound, these lists always include some head-scratchers, items in the "How the hell can we possibly do this?" category.

We're using a radical new landing design that has never been fully tested, because no tests on Earth can show how the design would fare on Mars. The needed actions are encoded in half a million lines of computer code, and there is zero margin of error.

If the landing isn't successful or the project comes to an early demise, we will spend months analyzing the failure to glean the lessons. But there will be those at high levels in government, along with prominent members of Congress, who will loudly lament that we should give up efforts to explore Mars and instead focus on other goals in space.

Every person in the control room knows some of the many ways our novel, intricate, never-fully-tested lander could fail. We all know the score for Mars missions so far is daunting. Of the thirty-nine spacecraft sent to

Mars, whether they were designed to fly past, to land, or to go into orbit, only nineteen so far have succeeded. Three of the successful missions were Soviet spacecraft, one was from the European Space Agency, and the rest were from the U.S. The only other attempts to join this exclusive club have been by Japan, with a 1998 effort that failed, and India, launching as I write this in early 2014. So the score is home team 19, Mars 20. That does not lend much confidence to the people who struggle to create missions like this one.

None of us wants to think of that on this night. Certainly none of us wants to think what it would mean for the Mars exploration program if we fail. So we sit there, watching and waiting. It's a surreal, sometimes scary experience.

Science team members from around the world, gathered in front of a giant screen in another building, take turns racing over to say hello, and race back to be certain they don't miss anything. Here in the control room, the peanuts are being passed around again—a tradition for these sessions when one of our spacecraft is arriving at its destination. A few of us, including me, are looking a little green from eating too many out of anxiety.

Another tradition, taken care of a little while ago, is the wakeup song, played about an hour before landing. Today it's Frank Sinatra crooning "All or Nothing at All."

Keith Comeaux, selected to be the flight director for the evening, will be working his way through the long checklist. Although by now the spacecraft is being directed by its own computer, no longer receiving commands from Earth, there are still many things to check and double-check. If something runs amiss in the next few hours or so before landing, Keith has a pile of contingency procedure documents that he can pull out and begin calling orders.

Each team member sits at his or her own display terminal, and I have, in addition, my own laptop containing detailed contingency plans for a bad day. If the worst should happen after landing, all eyes will turn to me, and I'll start calling meetings and kick-starting analyses in the desperate hope that a bad situation can be corrected. This is a position I fervently did not want to find myself in. We won't know of a problem until that 13.8 minutes after the bad has started. Any remedy I offered would take at least the same amount of time to transmit back to the lander, and by then it would probably be too late.

* * *

The single most challenging part of putting the rover on Mars, known in our business as EDL, for "entry, descent, and landing," is now taking place. Hundreds of people have worked on this one process. Computer simulations flashing on the screen follow the data flowing from the lander. The stream of events that is the hallmark of landing on Mars ticks off item by item.

A message arrives that the cruise stage has been successfully jettisoned. Years of work by more than a hundred engineers are over in a heartbeat. The spacecraft next does an about-face to aim its entry heat shield toward the planet. The most complex lander scheme in the history of spacecraft is getting its first all-up test 154 million miles (248 million km) away from home.

EDL team member Jody Davis sees that the event record she's looking for has appeared on her computer screen in the separate EDL war room. We all hear her voice over our headsets as she calls out, "Tango Delta nominal."

We each experience a surge of thrill. Tango and Delta are phonetic identifiers for the letters T and D, which the team uses to mean "touchdown." The "nominal" means "normal, fine, just as planned." Only insiders understand the phrase, used so that onlookers, reporters, and people in the global audience will not jump to a premature conclusion. It may mean that our years of effort will soon be successful—yet there is still plenty that can go wrong.

The next call comes quickly: "RIMU stable." The Rover Inertial Measurement Unit is signaling us that the rover is sitting on relatively level ground. It's not sliding down an embankment or being dragged across the surface.

On hearing that call, team member Brian Schratz has started counting to himself. If UHF radio signals are still arriving from Mars after the count of ten, it will confirm that the spacecraft itself has rocketed away from the rover as planned, to crash onto the Martian surface at a safe distance from the rover, Curiosity.

The project's EDL team leader, Adam Steltzner, is pointing at Brian waiting for that count. On reaching the count of ten, Brian announces, "UHF is good." Adam pokes Al Chen. Then what we are longing to hear: Al Chen, the EDL systems engineer, has been tasked with calling the play-by-play for the viewers on TV and the Internet. He makes the official announcement, calling out "Touchdown is confirmed! We're safe on Mars!"

The local time is 10:30 p.m., August 5, 2012; in New York, 1:30 a.m. August 6. Curiosity has come to a dead stop on the surface. Like a punctual guest coming to a dinner party, it has arrived exactly on schedule.

In a single unchoreographed moment we are all on our feet, thrilled and relieved, our hands thrust in the air. We are hugging each other and trading high-fives. I notice two of the guys trying to high-five, both so excited that they keep missing, until finally one grabs the other's arm, holds it steady, and slaps the now-motionless hand. The two, like several others, look as if they could break out in tears of grateful relief at any moment. I'm feeling the same. It really worked!

I walk into the room where invited guests have been watching and get a hug from longtime Hollywood actress and space fan June Lockhart, known for her roles as the mom in *Lassie* and of the *Lost in Space* family. I also get a big congratulatory smile from rapper will.i.am.

More than $2 billion has been gambled on this effort, while through the years our confidence of success in the mission eroded before gradually inching back up again. For so long it had seemed just too damned complex, with far too many paths to failure, leaving me at one point telling a team member, "It may be that nobody on Earth is smart enough and skilled enough to make this work." Yet we have stuck with it, seen it through, and found the ways to make it work, after all.

Soon the baton will change hands. Our ten years of work has been a prelude to what comes next, the purpose of the whole endeavor, when the scientists start their hunt for new knowledge about our sister planet. This will begin with exploring the area around the landing site on Gale Crater, and later, we hope, a trip to the foot of the mountain in the distance—places

FIGURE 2. MSL's control room staff at JPL, just after hearing, "We're safe on Mars." Rob Manning is seated at center left, wearing earphones. (Courtesy of NASA and JPL/Caltech)

that might hold the key to our own biological origins. We have gone there to uncover the ancient history of the planet, to discover if Mars ever had the ingredients to support life.

We're not sending robots to Mars. We're sending extensions of ourselves. These machines are *us*, and in visiting Mars, we may in time find that we are going home.

CHAPTER 1
The Red Planet Fixation

Spacemen have landed on Earth. Invaders from Mars are pouring off their space ships, spreading death and destruction. The creatures from Mars are "wriggling, glistening like wet leather, their faces indescribable." Americans are glued to their radios, listening to the alarming news bulletins and chilling eyewitness reports.

In 1939, this really happened. Not the Martian invasion, of course, but widespread panic among people who believed that a radio drama from actor and theatrical bad-boy Orson Welles was true, actually believing in the possibility that there were intelligent creatures on Mars, from a civilization so advanced that they could invade Earth.

The people taken in by what Welles and his team intended to be diverting entertainment seem not so far removed from people of earlier ages. Imagine you are living in the world of our distant Neolithic ancestors. You and your fellow tribespeople know how to gather food from the trees and plants, how to hunt, how to shelter yourselves in times of storm, how to procreate and raise your young. You tell the stories learned from your elders, stories that offer answers to the mysteries of life, of nature, weather, and the everchanging skies. In some versions, the sun has wings. In others, the sun crosses the sky on a chariot, or crosses the sky in a boat. The sun is reborn each morning. The moon is the daughter of the sun.

Across the span of time, tribes grow in size. Many settle down to farm. They build cities. They worship their gods, pleading to them for health, security, and prosperity, and perhaps for the defeat of their adversaries. Over eons, many come to worship a sole god—Yahweh, or Jesus, or Allah, or another deity. Finally they, our ancestors, begin to unlock nature's secrets. They turn from concocting fables to observing and studying, then laying down conclusions drawn from observed fact.

Those conclusions were largely askew. In the year 140 CE, the Greek astronomer Ptolemy published a complex mathematical model that accurately accounted for the apparent motions of the planets. For the next 1,300 years, his model kept science on the wrong foot, maintaining the doctrine that Earth is the center of the universe and all heavenly bodies circle around us. It was left to Copernicus in the 1500s to deduce that the planets travel in orbits around the sun. At last, humankind's curiosity about the visible sky was beginning to be based on something worthy of the label "science" instead of myths.

Around the same time Galileo Galilei became the first to examine Mars with the aid of a telescope, providing him the earliest close-up views, primitive though they were, of our nearest neighbor planet. It wasn't until the late 1600s that skywatchers began to note similarities between Earth and Mars. Christiaan Huygens observed that Mars was only a little over half the size of Earth, yet by what seemed like a remarkable coincidence had days almost exactly the same length as ours, later confirmed as a mere thirty-nine minutes longer.

Not long afterward, in the 1700s, German astronomer William Herschel kicked our thoughts about the possibility of life on Mars into high gear. He discovered another similarity between the two planets: Mars had what appeared to be polar ice caps, periodically growing and shrinking with the seasons just as on Earth. In the warmer seasons, dark areas were seen as possible vegetation.

That was a first step to speculation about the possibility of life on Mars. Odd as it now sounds, in the 1800s, the great German mathematician Johann Carl Friedrich Gauss actually suggested that huge geometric patterns be carved in the vast snowy regions of Siberia as signals to Martians who might be eyeing Earth for signs of life on our planet.

In time, we learned that Gauss was chasing shadows. Writers with creative minds have long teased the public imagination with tales of travel to the far reaches of the cosmos and encounters with non-Earthly yet intelligent life forms. We're fascinated with the possibility of discovering "what's really out there." From Orson Welles and his predecessors until today, science fiction writers and screenwriters have kept us beguiled, chilled, and frightened by imaginative stories inhabited by creatures from other planets.

Our most qualified thinkers on the subject give us reason to believe in the possibility. One of the most intriguing questions for all of humankind is this: Has life ever appeared anywhere else in the universe? Is it possible that intelligent life actually exists today elsewhere in the universe?

Scientists disagree, but for many the answer is all but certain. It is estimated that just the extent of the universe we can observe from Earth contains as many as 10^{24} stars. That's one septillion, the number 10 followed by 24 zeros. These stars are organized into galaxies. Recent measurements suggest that our Milky Way alone may contain 50 billion planets. About a billion of these are about the size of Earth and are inside the so-called habitable zone, where it's possible for liquid water to exist on the surface. Perhaps a tiny fraction of these billion planets would have been stable long enough for the development of complex multicellular life. Of these, a much smaller fraction could have contemplative, technologically advanced societies present today. Entire galaxies may have only a few planets harboring complex inquisitive life such as ours, yet the existence of billions upon billions of galaxies across the universe translates into a very large number of potential civilizations. And this is a very conservative estimate. In the 1970s, the late, revered Cornell astronomer Carl Sagan put the figure as high as one million potential civilizations in our galaxy alone.

Harvard professor Paul Horowitz recalls once telling a journalist, "Intelligent life in the universe? Guaranteed. Intelligent life in our galaxy? So overwhelmingly likely that I'd give you almost any odds you'd like." Today he says that he stands by that statement, with an amendment: "At the time, it may have been a bit daring, but in view of recent findings on the prevalence of habitable Earth-like extrasolar planets, it now seems, if anything, rather conservative."

Yet, even given billions of years, under what conditions do a bunch of chemicals, water, and various types of minerals evolve into living things?

Scientists have theories and are closing in on possible answers, but so far we simply don't know. We do know that it occurred surprisingly fast once Earth developed the right conditions very early on in its long history.

Although so far we have not detected life on Mars, a mythical observer seeing Mars and Earth from deep space early in their history might have labeled them nearly identical, neither more likely than the other to become a home where life might appear. Both Mars and Earth formed together along with the sun and the entire solar system about four and a half billion years ago.

Soon after the formation of Earth, a Mars-sized young planet slammed into our planet, creating a swarm of mountain-sized objects that continued to crash into our home for many millions of years. Other bits from that original impact coalesced to form our moon. Volcanic turmoil caused by these bombardments and the heat from radioactive rock spewed carbon dioxide vapor into the atmosphere. Finally our planet began to cool, the surface hardening into a crust. Water brought to Earth from the impact of millions of water-filled comets finally settled out of the hot atmosphere to become oceans.

What happened next is a story that scientists are only now unraveling. It starts perhaps four billion years ago. At the bottom of the oceans, acidic waters percolated down through cracks in the rocks to become laden with an alkaline-rich brew, then streamed back up inside hydrothermal vents, forming large towers of complex mineral structures. Tiny pores in these structures acted as "factories," tearing apart the nearly unbreakable bonds that hold together carbon-dioxide molecules. The result was a constant supply of organic molecules that would go on to become the bricks and mortar of life.

In time, the inner walls of some of these pores became coated with simple organic layers that formed protective yet porous bubbles around these organic molecule factories. Some of these bubbles escaped the pores to float with the ocean currents. Through a long series of fortunate accidents of nature, these little machines began to replicate. Within a few million years they developed tricks that allowed them to float across the oceans and adapt to new environments. These were very simple bacteria that could live on the simplest of chemical energy. Some of these simple nucleus-free bacteria called prokaryotes are still living today, thriving in the harshest conditions on Earth.

Even today, we live in a world that is not dominated by the plants, fish, animals, and insects that we know. An extraterrestrial studying Earth would report that the bulk of life here is made up of bacteria, and might mention as a footnote that some of those bacteria have congregated into large mobile colonies of cells, of which some have developed sentient behaviors that allow them to manipulate their world and even go off to explore others.

Steven Jay Gould, a Harvard professor, paleontologist, and leading thinker on evolutionary theory, wrote, "The most salient feature of life [on Earth] has been the stability of its bacterial mode from the beginning of the fossil record until today and, with little doubt, into all future time so long as the Earth endures. This is truly the 'age of bacteria'—as it was in the beginning, is now, and ever shall be. Bacteria represent the great success story of life's pathway." Elsewhere he noted that "during the course of life, the number of *E. coli* in the gut of each human being far exceeds the total number of people that now live and have ever lived."

Some of the facts about bacteria seem to defy common sense. A single gram of soil can contain on the order of forty million bacterial cells. From the earliest appearance of a single-cell life form, it took a few hundreds of millions of years and many more of those fortunate but unlikely accidents before the appearance of the first complex multicellular organisms, in what *New Yorker* writer Burkhard Bilger has called "life's triumphant march toward complexity."

Gould was skeptical that the march toward complexity, even if it occurs, would inevitably lead to advanced civilizations, but like the majority of scientists today he suspected that in the right conditions, single-cell life and the evolution it makes possible could well be inevitable. We know that rock ejected from large meteor impacts can travel across the solar system. After examining hundreds of meteors, scientists have confirmed that some come from Mars and other places in our solar system. Over the past few billion years, scientists believe that millions of tons of Earth rock containing many billions of cells have been ejected into space and spread to far away parts of the solar system. If even a very few of these cells survived the long journey frozen inside rocks, perhaps Earth life has populated other planets like Mars.

Chris McKay, a planetary scientist with the Space Science Division of NASA Ames Research Center, once gave an NPR interviewer what could be taken as a rationale for the project that is the story of this book: "Life may have started on Earth very quickly but purely by accident. Life may be easy to start under Earth-like conditions on any planet. These are questions that we'll never answer staying here on Earth. We've got to go look at another example. We've got to go see if it happened on Mars."

Is it possible that life began on Mars independent of Earth? It happened here. Why not there?

We go to Mars seeking answers to some of the great questions of our time:
- Was Mars ever habitable for simple single-cell organisms?
- Can we find evidence that past life once existed or perhaps even still exists on Mars?
- Can we find clues on Mars that will help us understand how life on Earth began?

One essential ingredient for life as we know it is water, liquid water. Water in the air, in rivers, ponds, puddles, lakes, oceans. Mars may have been warm and wet even before Earth was. Which leads us to other compelling questions:
- Is it possible that life began on Mars independent of Earth?
- Is it possible that life started on Earth and was transplanted to Mars?

And perhaps even more tantalizing:
- Is it possible that life started on Mars and was transplanted to Earth?

Some of these questions, I hope, will be answered in my lifetime and yours. The Mars Science Laboratory spacecraft has been designed to provide answers to some of them.

As we shall see, one has already been answered.

CHAPTER 2
Joining the Mars Adventure

Years ago, a young French farm boy was doing poorly in school. A wealthy neighbor took an interest in the youngster and asked him, "What do you want to do in life?" He answered, "To be a chef." It was a curious reply, since he had never even been inside a restaurant. Nonetheless, she arranged for him to be hired by a restaurant in a nearby city, Lyon, then famous as the home of some of the great restaurants in Europe.

Everyone on a French farm in that region was well experienced at plucking chickens and ducks, and the boy was soon the number-one plucker at the restaurant. Today that farm boy, Daniel Boulud, is the owner and chef of the restaurant Daniel in New York as well as a string of other restaurants in the city and elsewhere, and he is widely praised as one of the best chefs in the world.

Sometimes a person knows in some mysterious way at a very early age what he or she wants to do. I knew from the time I was twelve that I wanted to build robots and rocket ships and send them off to explore faraway parts of the universe. Like many kids of the 1960s and since, the work of NASA intrigued me from an early age. But unlike many other kids, I never outgrew this passion.

I spent part of my childhood living a Huckleberry Finn lifestyle on the islands and in the farm country of northwest Washington State. On our long, wet, windy winter nights, I was transfixed by the adventure of space and found it all so astonishing, barely able to believe people were actually building spaceships. Not only that, the spaceships took pictures that they could put on a radio beam and send down to Earth to be printed in the newspaper and delivered to our front door. I could hold the local newspaper in my hand and see pictures from space. It was a heady time to be a kid.

The funny part was that we knew so little about outer space. The Moon was clearly visible, but Mars was this blotchy red thing. Our classroom textbooks offered little about the outer planets beyond the simple observations that they were different colors, some were very big, and one looked like a coffee mug with two handles (Saturn with its rings). I read about Mars as a world possibly inhabited by some sorts of creatures. The idea of Martian cities of mysterious beings kept me brimming with excitement. The raw possibility of unseen worlds held the power to stoke my imagination.

One day the front page of our local paper showed real photos taken by a spacecraft flying past Mars. Later the front page showed photos taken by twin Viking landers that had reached the Martian surface, stirring images of what it might be like for *me* to be standing there on my own two feet.

But after all, there were no Martians, no aliens. The thought that Earth might be the exception after all, and that we are indeed alone, made me sad and kept me awake. I wanted to be one of the people to discover which version was true. But how?

For a kid like me, it wasn't going to be easy. My work in math and science in the classroom didn't exactly make me stand out. I was a mediocre student. But my curiosity about space exploration drove me to want to learn more. The teachers in the schools of my farming community said that I needed to do well in my studies before I could do anything technical.

One day, looking through the Time-Life book *The Scientist*, I came upon a picture of the graduating class of the California Institute of Technology. The text described Caltech graduates as "among the best" and said that these people would lead the coming technological revolution. I stared at the picture, studying the faces, longing to be like them.

When my family settled in the farming town of Burlington, Washington, I found the high school there a great place to learn the ropes of rural life: machine shop, auto shop, farm shop, small gas engines shop, plastic shop, welding shop, and lots of "ag" classes. After school and in the summertime, we learned the hard life on the farms and in farming industries. What we didn't learn was how to study. Given that background, I sent off my college applications with much trepidation.

It didn't seem possible that I could really become an engineer until I got to Whitman College, a small liberal arts school nestled in a comfortable corner of Walla Walla in the far southeast of Washington State. Whitman offered a "3–2" program where I could get two degrees—BA and BS—in five years of study: three at Whitman and two at Columbia—or, I was delighted to learn, at Caltech. But I needed to pull a decent grade point average. I had barely made it into Whitman, so qualifying for Caltech was going to be a challenge. Still, there is nothing like a good dose of fear to motivate a person. I studied hard, virtually living in the college library, and a few years later, I succeeded in transferring to Caltech.

I had my first brush with NASA and Caltech's Jet Propulsion Laboratory (JPL) within days of my arriving at the Caltech campus. I was assigned an adjunct professor of electrical engineering as my advisor. A gentle, elegantly dressed man with a New Zealand accent, he held forth in a narrow office loaded with magazines, newspapers, and electronic equipment, with his empty pipe sitting right next to a smoldering soldering iron. He generously spent hours with me giving advice on courses to take and areas to focus on. It was some time before it struck me that this man, William Pickering, was the father of JPL's planetary exploits, recently retired as JPL's director. Then I remembered seeing pictures of him in 1958, holding above his head in triumph a model of Explorer I, our country's first spacecraft. That this man was now my advisor was proof that I had come a long way from the islands of Puget Sound.

In my college years at Caltech, my classmates and I would drop everything to hang in front of a lecture hall TV monitor that piped fresh images as they arrived to Earth in real time from an Apollo mission or a Mars mission.

What else did those broadcasts show us? Images of the planetary scientists who had made Martian exploration possible, watching those same images with their mouths open, gawking at the screens, every bit as fascinated as we students were. At one of those viewing moments, I determined to become one of the people who made this all happen. I set my sights on JPL.

While I was still finishing my undergraduate work at Caltech, JPL offered me a part-time position as a draftsman to work on the schematics of the Galileo spacecraft, which was scheduled to be the first to fly into Jupiter's orbit. The thought that I would actually be playing even a small role in building a robot that would explore another planet was an absolutely fantastic notion. Just to be allowed to be at the place that spawned these machines was a dream come true for me. I would have accepted even if the job they offered me had been washing windows.

The origins of JPL, I already knew, were as unlikely as the unique work people do there. It had all started in the mid-1930s when some students from Caltech joined forces with a few amateur rocket enthusiasts to build experimental rocket motors right on campus. The tinkering went fine until one of their motors exploded. The rocketeers were invited to find another site for their experiments, and the place they settled on was an isolated area in the foothills of the nearby San Gabriel Mountains. By the time the 1930s gave way to the '40s and World War II, the group's accumulated experience with rockets led to contracts for advanced research and development with the Army.

When President Eisenhower signed the National Aeronautics and Space Act of 1958, the 'space' part of its title included exploring the planets. JPL was transferred from the U.S. Army's Ordinance Department to the new NASA, but unlike the Goddard Space Center, the Johnson Space Center, and the other NASA centers, JPL became the agency's only contractor-operated center. (The actual arrangement is somewhat complicated: JPL is managed for NASA by Caltech. JPL staffers are Caltech employees, with our paychecks reading "California Institute of Technology.")

Today JPL, though of course much expanded, is still nestled in its original foothills location, just a few feet away from the very spot where those early rockets were fired. The setting is rural, with a famous riding academy next door. Walk across the Lab's campus and you'll pass grazing mule deer; born on the grounds, they are free to wander off into the wild but instead choose to remain and live their lives at JPL, presumably never noticing how much their presence surprises visitors. The atmosphere at JPL is cerebral, yet, unlike what you would find at most of the great universities, there is a restlessness here, as if each person you pass is struggling with a problem never encountered before and for which there is no apparent solution. Many of those people, then and now, are grappling with some problem involved with the exploration of Mars.

When I arrived at JPL, I found myself inside the very world I had envied more than a decade earlier: I had become part of the space team. Very cool. I was suddenly around the people who invented robots that went into outer space—including some of the very people I had been so awed by as a kid, and I found myself at times in a position to learn from a number of those personal heroes. For a young person eager to learn, it doesn't get much better than that.

Once I had my diploma in hand, JPL changed my status from draftsman to engineer, but my role as an engineer was slow going. My first job was as an apprentice electronics tester, helping run tests on what would become the brains of the Galileo spacecraft. I quickly discovered that building spacecraft included many extremely tedious jobs. After Galileo, I worked on Magellan (to Venus) and Cassini (to Saturn), becoming expert in the design of spacecraft computers, computer memory, computer architectures, and fault-tolerant systems.

In 1993, after thirteen years at JPL, my career took a sudden leap forward. Brian Muirhead, the most inspiring and level-headed spacecraft leader I have ever met, had recently been named spacecraft manager for a funky little mission to Mars called Pathfinder. We had a conversation in which he explained that he was a master of mechanical systems but had not had much experience with electronics. He asked if I would join him as his chief engineer and deputy spacecraft manager.

This was a whole new world. I could not believe my luck. It had been many years since anyone had landed anything on another celestial body. In addition, Pathfinder would carry Sojourner, the first rover ever to land on Mars. That we were expected to do it on the cheap made it seem an even more exciting challenge: We would have to make it as simple as possible.

In addition to my original responsibilities, I was soon handed the role of leading the creation for Pathfinder's entry, descent, landing, a subject none of us had any experience with. I ended up spending 90 percent of my time solving the problems of EDL. I lived through one of the most intense experiences that a spacecraft designer could undergo. Our tiny team of dedicated young people held on to a crazy notion that we could design a spacecraft, a heat shield, a supersonic parachute, solid rockets, plus something never done before, a risky first—giant airbags that would cushion the first rover ever to reach Mars—and do all this for no more than the cost of a major motion picture. Pathfinder was among the first NASA missions to employ a culture of what the then-administrator of NASA termed "faster, better, cheaper." With a willingness to take risks, we learned as best we could how to build and test the pieces and how to put them together into something that had a good chance of working.

It did work. Our efforts paid off on July 4, 1997, when Pathfinder made the first landing on Mars in twenty years. Its tiny payload, Sojourner rover, made the first tracks on the surface of another planet in human history.

The previous year, my teammates and I had learned from a NASA email about an upcoming major announcement concerning Mars. Together, we watched the televised press briefing in wonder. President Clinton introduced it with these words: "I am determined that the American space program will put its full intellectual power and technological prowess behind the search for further evidence of life on Mars."

At that press conference, NASA scientists announced that they had discovered small carbonate globules inside a Mars meteorite that had been picked up in 1984 from atop the snow at the foot of the Allan Hills in Antarctica. They concluded that these tiny structures were evidence of microscopic Martian bacteria.

The rock had been formed very early in Mars history, some four billion years ago, less than a billion years after our solar system was formed. About 15 million years ago, well after the end of the dinosaurs but well before *Homo sapiens*, a large impact on Mars ejected this old rock into outer space, where it spent most of the rest of it days circling the sun. At the end of the last ice age, about 13,000 years ago, it crashed into Earth.

After years of study, most scientists agreed that the evidence found in the meteorite was inconclusive: there were non-biological explanations that might account for the observations. But by then the original announcement had set in motion a huge desire within NASA and the scientific community to try a "sample return"—a space mission that would involve landing on Mars, gathering samples of soil and rock, and somehow returning them to Earth. Enthusiasm was running high. Mars exploration was on a roll.

* * *

In September 1999, a visit to Lockheed put me at the Embassy Suites in Denver on the morning that the NASA/JPL spacecraft called the Mars Climate Orbiter was scheduled to reach Mars. I woke at six in the morning, annoyed that I had slept through the arrival time, and turned on the television. Instead of smiles, I saw long faces and incredulous looks. There was the Lockheed program manager I had been talking to the day before, explaining that no message had arrived from the Climate Orbiter to confirm its arrival in orbit.

I raced to dress and hurried to the lobby, knowing that several others from JPL were staying in the same hotel. In the breakfast area, I found JPL's George Pace and Roger Gibbs. The three of us stared at the TV monitor, appalled. A CNN reporter was interviewing Lockheed and JPL managers, people we all knew well, about what appeared to be an improbable failure. We were stunned, almost disbelieving.

What was harder for us to take in was the fact that the project managers at JPL were pronouncing the mission a failure only four hours after it was supposed to have arrived. That never happens. When we lose contact with a spacecraft, we try for days or even weeks to regain contact. Some big calamity must have happened that the team wasn't acknowledging publicly yet. What had failed?

With my stomach in knots, I grabbed the next flight home. Since it was my daughter Caline's second birthday, my wife Dominique and I headed to the Los Angeles Zoo with her as soon as I arrived. But so much was rattling in my head that I could barely focus. My friend and JPL colleague Wayne Lee called me just as we parked at the zoo, saying, "Rob, you won't believe this. The navigators tell me the Orbiter may have hit Mars! They don't know yet how it could happen, but there are a lot of very unhappy people here."

I was astonished. If there is one thing JPL has been doing right for decades, it's navigating to the planets with mind-boggling precision. On Pathfinder, we had managed to target a bull's-eye above the Mars atmosphere that was only a few kilometers wide. It was like golfing on a course from Los Angeles to New York and sinking the ball on the fourth stroke—with the hole moving at several thousand miles an hour.

Clearly there had been some sort of unprecedented navigation error. Instead of flying well above the Martian atmosphere as it fired its main engines to be captured into Mars orbit, the spacecraft had crashed into the atmosphere at more than 12,000 miles (19,000 km) per hour. The Mars Climate Orbiter had become a lander.

At JPL, arguments were flying and tempers were flaring. How could this happen?

Within a day, the answer had been found. The spacecraft had fallen victim to the one of the simplest possible mistakes. It was equipped with tiny thrusters to keep its solar panel facing the sun. But knowing that tiny nudges from these thrusters would also ever so slightly nudge the spacecraft off course, the team's navigators asked that the spacecraft team keep track of and report the small forces of the thruster firings as the spacecraft journeyed to Mars. The reports were to be given in SI (metric) units called Newtons. Because of a programming oversight in the software used by the spacecraft team, the force data had been mistakenly delivered to the navigation team in English units, pounds. One pound is more than four times larger than a Newton.

When accumulated over months, these small forces ended up pushing the spacecraft off course, nudging it much closer to Mars and putting it on a collision course with the top of the Martian atmosphere. As it made its arrival from deep space, it fired its main engine to enter into its first orbit about Mars but due to that programming error was already too close. Without a heat shield to protect it against the soaring temperature buildup as it entered and plowed through the Martian atmosphere, little Mars Climate Orbiter must have broken up like a meteor. What little of it survived the heating would fall destroyed to the surface.

The trick for successfully building complex one-of-a-kind machines where hundreds of thousands of things need to be done exactly right is to *expect* mistakes. Humans make mistakes. We all do. For all spacecraft designers, Mars Climate Orbiter became a lesson to remember that is essential for mission success: Test independently. The real mistake made on Mars Climate Orbiter was not the English-to-metric error, but an error in not checking thoroughly enough for errors.

With the Climate Orbiter gone, the team at JPL and Lockheed was devastated, and not just for the obvious reason. For the most part, the same people who had conceived and built the Orbiter had done the same for the second spacecraft of the pair, the Mars Polar Lander, for which they had used much of the same technology. The Lander was now only weeks away from Mars. The team needed to put the Orbiter failure behind them quickly and focus on making sure the Lander would not suffer a similar fate. There would be time later to mourn the loss of the Climate Orbiter.

Both Climate Orbiter and Polar Lander had been designed sparingly under a tight fixed-price arrangement that had been worked out among Lockheed, JPL, and NASA. Only elements absolutely essential for landing and doing the science were on board.

I had recently been appointed to a new position as the Mars Surveyor Program's first chief engineer. As one of my duties, I was to sit on engineering review boards for the Mars missions then being developed. The previous April, I had attended a review of the Lander's EDL design. I was impressed at how much the team of people at Lockheed Martin Astronautics, working under contract to JPL, had accomplished with so little money.

But one item troubled me: As the spacecraft approached for landing, the radio was to be turned off. If something went wrong, we would have no information transmitted from the Lander to let us know what had happened, no data that could provide valuable clues to prevent the same fate on later spacecraft.

When I pushed the team members about keeping the radio on during the short, vital period, only a matter of tens of seconds, they explained the issue. Those radio transmissions might interfere with the signals of the radar altimeter that provides the essential "how far to the surface" data fundamental to a safe landing. The only way to prove the radio could be safely used would be to conduct a series of tests. But this project was on a fixed-price contract; there simple was not enough money remaining to do anything beyond getting the spacecraft ready for launch.

Back at JPL, I pleaded with my boss, the Mars Surveyor Program manager, about getting extra funding from NASA to cover costs for the extra testing. At the time, funding for Mars projects was cobbled together from a number of funding sources within NASA. He explained that there was no one person he could turn to for the additional money.

Friday, December 3, 1999. Too many of us were crowded into a small control room at JPL, not much bigger than the dining room of a small apartment, surrounded on two sides by glass-windowed walls for reporters and others to peer through. It was landing day for the Mars Polar Lander. Compared with the disciplined and practiced landing-day events of later Mars missions, there was an air of chaos to these proceedings. There were few visual displays that would give watchers of NASA-TV or the nascent Internet a clear sense of what was going on.

The head of NASA, Dan Goldin, was sitting next to Ed Stone, JPL's director. Inside the control room was the Polar Lander mission manager, my old friend from the Pathfinder project, Sam Thurman. Over Sam's shoulder was a radio spectrum analyzer display that showed the signal coming from the spacecraft. As expected, seconds before cruise stage separation and about ten minutes before entering the top of the atmosphere, the radio was turned off, and the signal disappeared. It would be about fifteen minutes until landing.

A quarter-hour later, after the lander should have touched down, we began watching for a tiny spike of the radio's carrier frequency that would pop up on the display monitors, indicating that the lander had settled onto Mars.

I stood up on my toes, craning over Sam's shoulder, watching for that little spike to appear.

Five . . . four . . . three . . . two . . . one . . . and . . .

No spike. Nothing.

I was suddenly gripped by an attack of nausea. Slowly, hoping others wouldn't notice, I inched my way to a back door and slipped out.

The team continued into the fourth day, hoping a signal would appear. They continued to hope but probably suspected the same thing I did: it wouldn't. Yet they frustrated themselves and millions of followers by reporting daily that they expected the signal to return at any minute.

Later searches by various Mars orbiters would never find any sign of the lander. No wreckage, no artificial crater. We spent a decade hunting for it. The lander had simply vanished, likely buried under Martian CO_2 dust and rocks that settled over it.

There seemed to be a consensus that the "faster, better, cheaper" model, which had worked so well on Mars Pathfinder, was a root cause for both failures. A dark joke went around the corridors of JPL: "Faster, better, cheaper—pick two."

With the failures of the Climate Orbiter and the Polar Lander, the reputations of JPL and NASA had suffered a serious blow. A cloud of uncertainty hung over us. We—the "rocket scientists" of JPL, considered the world's leading experts in building and flying spacecraft—found ourselves now left with no plans for another mission to the surface of Mars. We had to go back to the drawing board.

The disasters made us fodder for late-night television hosts. "It proves," Jay Leno quipped to his large national audience, "that you don't need to be a rocket scientist to be a rocket scientist." The imagined sounds of the nation laughing rang in our ears.

CHAPTER 3
Avoiding Splat

As a result of the failure, the hardware for a follow-up project based on Mars Polar Lander was mothballed, and so were far more ambitious plans for a Mars sample return rover or a roving laboratory mission. These missions would have required landing rovers far larger and more complex than anything we had done before. We came to realize that we did not know how to land *anything* on Mars reliably, let alone something large. It was quite possible that Polar Lander did not fail its landing, but that we failed Polar Lander. We really did not have a clue about what the spacecraft's landing site looked like, since the images from space were not crisp enough to reveal the level of detail we needed. Perhaps it landed on a steep slope. Perhaps it fell into a crack. Perhaps it was the victim of a design fault in the machine. There were certainly plenty of other possible explanations. We would not likely ever know.

NASA wanted to get back on its feet by regaining public confidence. The only way to do that was with a Mars project that would be such a success it would make people forget about the failures.

Within weeks, JPL formed a team to think about new ideas for delivering a large rover safely onto the Martian surface. Though we did not know it at the time, we were taking the first step to designing the landing technology for what would become the Mars Science Laboratory spacecraft and its rover,

Curiosity. We were being given the challenge of revisiting the whole architecture for how to land on Mars without going splat.

The Wright brothers mastered flight in 1902, including the art of landing. American astronauts first landed on the moon in 1969. I am often asked why landing on Mars is so much harder than landing on the moon or on Earth.

To land on the moon, the astronauts entered lunar orbit and fired retrorockets aimed more or less opposite to their direction of travel. As their spacecraft slowed, it descended toward the surface. The landing isn't trivial, but it's reasonably straightforward.

To bring a lander back to Earth, retrorockets aren't needed, because Earth has an atmosphere. Most Earth landers can eliminate more than 99 percent of the speed of orbit simply by slowing down with a heat shield. For the last 1 percent, we can use parachutes (as did Soyuz) or wings (as did the Space Shuttle).

Mars is neither like the moon nor like the Earth, but instead is annoyingly in between. It has too much atmosphere to land as we do on the moon and not enough to land as we do on Earth. The thickness of the Martian atmosphere at the surface is similar to what a mountaineer on Earth would feel if standing on top of a mountain 130,000 feet (40,000 m) high—four and a half times higher than Mount Everest. At that altitude, the Space Shuttle is still screaming along at over 4,000 miles per hour (1,800 m/s). How do you slow down quickly enough that by the time that you reach the ground you're not still going over 1,000 miles (1,600 km) an hour?

That's the problem we face when designing our Mars landers. It's the reason our machines are such Rube Goldbergesque contraptions and why the seven minutes of entry, descent, and landing are so terrifying. We have to combine all of the tricks we use to land on Earth (heat shields, parachutes) with the techniques we use to land on the moon (retrorockets, airbags), among many others.

The planet is riddled with huge boulders, ravines, canyons, craters, and peaks. The best images we had of the Martian surface did not provide enough detail to see any but the most dangerous boulders, yet everything we knew warned that nearly any place we landed would have rocks and slopes. There are few regions on Mars that are free of vicious landing hazards.

Making matters worse, the Martian atmosphere also varies widely from place to place and day to day, creating a huge uncertainty in where the spacecraft will finally come to land. It could be anywhere within about 500 square miles (805 sq km)—an area about the size of Los Angeles. All of our Mars landers have had large landing areas called "landing ellipses." The first US Mars landers, Viking I and II, had a landing ellipse that was 175 miles (280 km) long and 160 miles (100 km) across. This uncertainty has forced us to find safe landing areas that are huge. Without some sort of autopilot to guide our vehicle through that uncertain atmosphere, we would have to struggle to find safe areas on Mars.

It began to look as if any Mars landing would require some kind of robotic Neil Armstrong looking out the window to guide our lander to a safe spot. But, of course, no such technology existed.

As it turned out, Armstrong and the techniques of the Apollo man-on-the-moon era offered what became one of the most important elements that transformed our efforts from "wouldn't be great if we could . . ." into a viable project.

Back in the 1960s, NASA engineers had needed to make sure that the astronauts returning from the moon would land reasonably close to the recovery ships in the Pacific Ocean. The spacecraft would somehow have to be steerable during reentry.

The pioneering engineers at NASA's Johnson Space Center developed a solution that would turn the entry space capsule into a hypersonic aircraft. They conceived a way to shift the weight inside the capsule so that when approaching the Earth, the heat shield would be tilted slightly downward—enough to provide a little bit of lift.

But big, round, flat heat shields do not have ailerons or wing flaps like airplanes that would allow the pilot to bank the wings to the left or right. The NASA engineers came up with the idea of adapting the small thrusters designed for keeping the capsule stabilized in outer space. They recognized that the thrusters, angled on either side of the capsule, could be brought back into use during reentry. By firing the appropriate combination of thrusters, the capsule could be forced to roll to the left or right, steering toward the desired touchdown point.

This technique, which came to be called "entry guidance," continued to be used for bringing astronauts and cosmonauts back to Earth, so it was in the bag of tricks that we naturally considered. Still, due to the thinner Mars atmosphere, it would have to be used much closer to the surface than on Earth. All we would need was to add the same type of thrusters to our Mars entry capsule. We could use the on-board computer to fire those thrusters in bursts to bank the capsule to the right and left, creating long, shallow S-curves. If the entry guidance software detected that the lander was overshooting the landing spot, it would command larger S-turns; if undershooting, smaller S-turns or none at all.

What's more, JPL studies in the 1990s pointed the way toward basing our Mars entry guidance software on the same design that had been used to return Neil Armstrong and the other Apollo astronauts back to Earth.

My teammates and I tried this idea out on an old Pathfinder teammate and Mars scientist, Matt Golombek, who is JPL's resident expert on picking safe landing sites on Mars. For years he had been struggling to find sites that were safe as well as promising enough to please the scientists. Unfortunately, the more we learned about the rough surface of Mars, the more challenging his task had become. But with this new guidance entry concept, instead of having to aim for a landing somewhere within an area more or less the size of greater Los Angeles, we could aim to land within an area no larger than downtown Manhattan.

Matt's enthusiasm was heartening. He immediately understood how by improving the landing accuracy, his job of finding scientifically interesting places that were also safe to land at would be vastly easier. He could even envision landing inside large craters or snuggling up a landing site at the base of a mountain. The benefits for science were too great to ignore.

This technology of the 1960s became the cornerstone for the spacecraft we would soon be planning.

When it came to the question of designing a lander that could settle safely onto the Martian surface, we had only a pair of landing techniques in our tool kit: legged landers and airbag landers. Over the previous eight years, we had struggled with both, and we knew that both had Achilles' heels.

One overwhelming challenge we faced and would likely face on future Mars projects was this: How do you put a much bigger spacecraft down

FIGURE 3. An artist's rendering of a legged lander, used on the Phoenix Mars mission. (Courtesy of NASA and JPL/Caltech)

FIGURE 4. An artist's rendering of an airbag lander, designed to bounce onto the surface, used on the Spirit and Opportunity missions. (Courtesy of NASA, JPL/Caltech, and Dan Maas)

safely? Our two landing techniques combined with our tried-and-true Viking landing design worked for landers the size of a dining-room table, but could they be scaled up for missions that needed something larger? We were talking about a legged lander—one with three side-stretched legs much like the twin Viking landers of the 1970s. Although I understood EDL from studying Viking and from leading the EDL team on Mars Pathfinder, touching down with a much bigger legged lander on Mars presented challenges that looked daunting.

On a legged lander, the crushable legs were designed to cushion the landing. No other part of the lander would touch the ground. Though we thought of them as "soft landers," they still hit the surface at a brisk speed. Any slower and there would be a risk that the force of the supersonic plume from the descent engines would have time to dig holes into the ground, which the lander's legs could fall into.

If a lander came down where the ground was too tilted, it could tip and roll over. If there were excessively large rocks, they could puncture the lander body or even one of the pressurized fuel tanks, possibly causing an explosion. The images we had of the Mars surface taken by the Viking orbiters and Mars Global Surveyor did not provide enough detail to show us all the small but dangerous boulders on the Martian surface. And if the lander software were too slow at detecting the moment of first contact with the surface and failed to turn off the engines fast enough, the lander could bounce and flip over.

If this lander had to be designed to deliver a rover perhaps as large as a golf cart, we would need to provide some way of getting the rover off the lander's top deck under all of our expected rock and slope conditions. Given the height of a legged lander that could deliver a rover, we realized that we would need long, sturdy, and heavy ten-foot (3 m) ramps for the rover to drive down.

At the same time, the big rover-delivery lander we had been studying was starting to look huge, top heavy, and weighing much more than a launch vehicle could handle. The harder we tried to make it all work, the heavier and more complex it got, and the more frustrating.

Of course, there was another landing technique from our toolkit: How about going back to the Mars Pathfinder airbag idea and let our large rover bounce onto Mars surrounded by an airbag cocoon? In 1997, we had successfully landed Pathfinder on large rocks and rough slopes using airbags. Instead of falling away from the entry capsule and landing on legs, for Pathfinder we lowered the lander on a rope, hanging under the parachute while falling through the thin Martian atmosphere at more than 150 mph (240 kph). We kept the lander suspended under the parachute until the last possible second. At about 500 feet (150 m) above the ground the airbags were inflated and three big solid rockets fired for a couple of seconds. The lander came to almost a dead stop some 50 feet (15 m) up. From there it fell free and bounced over the terrain for the next minute or so until coming to rest.

In principle, this simple design concept should also work for a larger lander that contained a much bigger rover. But we found that even the strongest bulletproof fabric could not survive a sharp rock on Mars with a heavier payload at high impact speeds. A rover much bigger than the size of the Pathfinder lander just didn't seem to be in the cards.

In early 2000, just weeks after the loss of Polar Lander, Brian Muirhead, who had brought me into the Mars Pathfinder project, was given the assignment of creating a study team to search for solutions to safe entry, descent, and landing. I joined in, and we gave ourselves three months to use our best people and experiences from Viking, Pathfinder, and Polar Lander to look again at the whole Mars EDL concept from top to bottom. Perhaps by taking the lessons we had learned from the earlier projects, we could come up with ideas that might help. Brian's organization chart for our study team was drawn with a circle around each subtopic. There were so many circles that the chart looked like a tub full of soap bubbles, and we were soon calling ourselves "the bubbleheads." We dove in to address a wide range of EDL questions, among them these:

- Were there more reliable ways of ensuring that a lander could be guided to the desired point of entry at the top of the Martian atmosphere?
- What kind of information and imagery would we need from future Mars orbiter spacecraft to help us pick places on Mars that would be safe enough to land?
- Could we figure out how to guide our spacecraft to a smaller landing target area?

- How could we provide for slowing down a heavy spacecraft: Larger parachutes? Multiple parachutes? Larger rockets?
- How do we guide our landers so that they could land within an area the size of a city block instead of a large county? Could we keep track of where the lander was in relation to where we wanted it to land? Could the lander see where it was on its map, as an astronaut would? Or would we need to place a beacon on Mars for future landers to home in on?
- How might a lander be able to see big rocks and steep slopes, and avoid them?
- How rugged do we need to make the lander? How could we land a big rover more gently than Polar Lander and yet make it rugged enough to handle slopes and rocks?
- If the lander needed to be high up off the ground to avoid rocks, how would the rover get off the lander?

I took on the job of looking across all of these bubbles to make sense out of them and see if there was a common thread or a common architecture that would glue them together.

Sometimes it's important to remind ourselves that there are no stupid questions, only stupid answers. We encouraged the teams to ask, ask, and ask. When we didn't know, we would say so. Arrogance has no place in this work. No matter how smart you think you are, if your Mars lander didn't work, you were probably just not smart enough, and if it did, well, maybe you just got lucky.

In the short and highly productive out-of-the-box thinking of the bubble teams, we came up with some important notions:

- The lander would need to be able to land on slopes perhaps as steep as 25–30 degrees, and on rocks at least as high as 1.6 feet (0.5 m).
- We would need a new Mars orbiter—our version of a spy satellite—that could reveal the location of rocks this size and larger, to aid in eliminating hazardous landing sites.
- We would need to devise a method that would ensure landing within a much smaller, more precisely defined area.
- The lander would need to be rugged for rocks but not so tall that it would put the rover high above the ground and away from rocks using legs. If

after landing the rover itself was still some 3–6 feet (c. 1–2 m) above the surface, in effect it would not have landed yet.

That last one was a tough requirement. We had only recently been discovering that Mars is full of tall rocks, and now had to recognize that our previous landers were not high enough to land without possible damage to the rover. My old friend, airbag designer Tom Rivellini, suggested to the team a change in the shape of the lander. He proposed a lander in the shape of a flat, star-shaped pallet that did not need to be suspended above the ground on legs. Instead of clearing the rocks, it would make contact with them and the surface. Crushable foam could be placed under the belly to keep the fuel tanks from rock puncture. It would have flexible "outrigger" legs that would splay out to keep it from tilting on steep slopes. Once on the ground, the rover would be able to drive directly off the pallet without ramps. These concepts gave rise to what would come to be called the "pallet lander."

The challenge would be to design the lander structure such that it could somehow contain and protect the landing propulsion system—throttled engines, fuel tanks, and electronics—against damage when landing on rough, steep, and rocky surfaces.

As the bubbleheads proposed every wild idea for landing and roving Mars they could think of, I began to consider how the ideas might be related to each other. What we needed was something like a "family tree" diagram, branching out to include all of the options starting with the approach to Mars from deep space, through entry, parachute inflation, descent, the terminal phase near the ground, touchdown, and finally egress of the rover. At each phase of the landing process, the family-tree diagram would broaden as options that people had thought of so far branched out. It would be what I called a "taxonomy" of the EDL design. I thought it might reveal pathways we had not yet explored.

Near the end of January 2000, I started on a big sheet of paper to see what this would look like. After spending a morning enumerating all possible approach and entry branches, I went on to the descent part of the tree. Single supersonic parachute, clustered-chutes, two-stage parachutes, liquid versus solid rockets—whether it made sense or not, it went onto the tree.

For Pathfinder, we had used a design vastly different from all the earlier ones, with propulsive solid rockets that brought the lander to a full stop 50 feet (15.2 m) above the surface. Why hadn't we instead timed the solid rockets

FIGURE 5. An artist's rendering of the pallet lander originally considered for MSL. (Courtesy of NASA and JPL/Caltech)

so that the lander came to a stop just inches above the ground? I would have loved to do exactly that; it would have made the airbag design and test job a lot easier. But we just couldn't get that approach to work. With uncontrolled solid rockets and a parachute pulling this way and that, we just did not have the control of the lander's speed and position above the ground to permit a close-to-the-ground stop.

The designs for other missions, in contrast, had used throttled engines (Viking) or pulsed engines (Polar Lander) that controlled the speed and position of the lander hanging below. Could those two approaches be combined? I put this option and others similar to it on my taxonomy chart.

To review the possibilities, I got together with fellow bubblehead Steve Jolly from Lockheed and with Dara Sabahi, my old friend and JPL's chief mechanical engineer, who is blessed with a fantastic sense of technology rights and wrongs. I trusted their instincts, knowing they would not be put off by silly combinations of ideas.

We gathered at a long table in the windowless conference room we had been using on the fourth floor of JPL's Building 264, and went through the taxonomy drawings I had made. When we got to my tree with landing architecture options, I pointed out one particularly kooky notion of a system that combined two elements: the precision-controlled propulsion design concept from the Viking and Polar lander projects, and the suspended propulsion configuration of Mars Pathfinder. If a Pathfinder-like lander or rover had more precise control of its rockets, this propulsion system as part of an "entry and descent stage" might be able to lower the lander closer to the ground and slow its descent enough to make a super-soft touchdown.

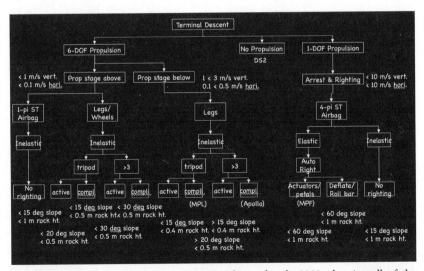

FIGURE 6. Rob Manning's design taxonomy chart of early 2000, showing all of the different possibilities for landing that the engineering team could conceive at the time. (Courtesy of NASA and JPL/Caltech)

I pointed out to Dara, "We could make the airbags and the surrounding lander a *lot* smaller."

"In fact," Dara asked, "why do we even need airbags and a lander?"

That made us all start thinking, "Could we land a rover directly on its wheels?"

Steve suggested, "This 'descent stage' could also do the entry guidance. It could be a general-purpose landing system that delivers packages to the surface."

The three of us realized together we were on to something. Maybe.

Or maybe not.

Either way, our new term "descent stage" would eventually become part of the vocabulary of MSL's landing technology. We knew that, like every EDL idea we had ever worked on, it would have some sort of downside, likely a major one.

Later that day, we presented the idea in a whiteboard session with a few of our teammates, including Miguel San Martin, the best guidance-and-control engineer and architect at JPL. He's a good friend, a passionate Argentinian armed with a keen sense of humor. (I especially remember his comment in one meeting that "If we try to add one more layer of detail to the Excel schedule, there won't be enough colors in the rainbow to handle it.") To my relief, Miguel was eager to consider it. Instead of arguing against the whole idea, he spotted a flaw and leaped right in.

"No," he said. "You drew the ropes wrong!"

I didn't know if I should feel happy that Miguel saw only one problem or insulted for being an incompetent artist when it came to drawing ropes.

"Put them in like *this*!" he said, redrawing the picture.

I had drawn three ropes going up from a single point on the rover to three separate places on the descent stage. Miguel pointed out that at this point in the landing, the descent stage and rover would be traveling straight down. If the rover was hit by wind or landed on a tilt, my configuration would force the descent stage to tilt. That would create a *horizontal* component of thrust, and the rover would touch down while being dragged across the surface.

Miguel explained why the ropes needed to go up from the rover to a single point smack in the middle of the descent stage. With his configuration, a puff of wind or a touchdown where the ground wasn't level would still cause the

FIGURE 7. Rob Manning's configuration of bridle "ropes" (at left); Miguel San Martin's version (at right). In the first version, tilting of the rover would cause tilting of the descent stage, which could cause the rover to be dragged across the surface, while with the second version, any tilting of the rover would not cause the descent stage to tilt. (Courtesy of NASA and JPL/Caltech)

rover to swing away from vertical, but this would not cause the descent stage to tilt. It would continue its controlled vertical descent.

Of course! Simple, but brilliant.

The next day, we brought this idea to the larger group to mull over. Joel Krajewski had drawn a cartoon of what this EDL and landing might look like. There was a lot of discussion both for and against. I could sense the development of two camps within the bubbleheads. One group was as cautiously enthusiastic as I was and another larger group thought the idea was too far out there. Several of them argued that controlling something swinging on the end of a rope was going to be next to impossible. We had to admit that none of us had ever heard of a solution for controlling a heavy object swinging freely under rocket power.

In particular, my old Pathfinder EDL colleague Sam Thurman was skeptical that using throttled rockets would be sufficient to control the touchdown speed of a rover hanging below. He was concerned about wind, oscillations, and other pendulum-like dynamics. "Rob," he said, "while I have to admit it might work, as of today we do not know how to control a hanging rover with the precision needed to land directly on its wheels. This is a pendulum. This is not a controlled system—it's going to swing. It could be uncontrollable."

Sam was right. We did not know if we could ever find a control scheme that would prevent the rover from swinging. I felt deflated. Reluctantly, Dara and I had to agree that, while it appeared promising, it was no more promising than the idea of using oversized airbags. We simply could not say with full confidence that this rover-on-a-rope architecture would pan out. Ultimately the team decided that we should endorse an architecture that appeared to have a better chance of success, and in their view, that architecture was Tom Rivellini's new pallet lander concept. In his new configuration, the rover would sit on a crushable and deformable pallet.

That was the theory, and despite the recent failure of Mars Polar Lander, we were choosing not to deviate far from the design of that ill-fated spacecraft.

The rover-on-a-rope idea that we had thought was so promising was now dead.

In the spring of 2000, Scott Hubbard, appointed to the new position at NASA headquarters that we would come to call the "Mars Czar," had to pull a program together for the 2003 and 2005 launch opportunities—dates dictated by the fact that Mars and Earth are in close proximity only once every twenty-six months. There wasn't a single project in the pipeline for either. Especially for the first one, only three years away, he would have to get something rolling quickly.

We received word that Hubbard, along with an independent review team, would hold a "shootout," more formally called the "Mars Options Assessment Review," where they would hear oral presentations for project ideas. JPL, Lockheed, and other organizations were told to generate mission concepts.

A few days later, Mark Adler popped into my office. He said, "Rob, there's only one way we can get a rover ready to launch to Mars by June 2003. We have to use Pathfinder." With only a little more explanation, I understood what he meant. His idea was to reuse the Pathfinder design including its airbag landing system. In place of its little Sojourner rover, we would convert Pathfinder's lander into a rover that would be folded up tight inside. Without Sojourner, and without all of Pathfinder's innards, would this leave enough room to cram a larger rover? Sojourner had been like a dog on a leash, linked to the lander by a short-range radio. It couldn't perform any useful science beyond radio range from its mother lander, no more than a few tens of meters, because it had no way to talk to Earth except by feeding its data to the lander.

If we *could* figure out how to stuff a larger rover inside Pathfinder's pyramid-shaped lander, what we would be offering was an idea for a rover that would revolutionize the exploration of Mars—a rover that could, well, *rove*. Since it wouldn't have to be tethered by a radio or a cable to its lander, it could be free to drive in any direction to sites the scientists thought looked promising—limited only by the usual trio of landscape villains: rocks, gullies, and steep inclines.

The design would also allow for a vastly greater variety of landing sites: Because the rover would be able to travel significant distances, we wouldn't have to land in exactly the area that the scientists wanted to explore.

All of the lofty roving dreams seemed to miss the obvious point that humans had really explored only three house-size spots on Mars. The planet is a bit bigger than that. In fact, the surface area on Mars is about equal to the land area here on Earth.

The team that Mark and I put together took less than two weeks to develop a design that we thought might work for the concept we had come up with. Mark took on the challenge of preparing the twenty-minute presentation. The night before, he was still a long way from finishing, but said he'd work all night if necessary.

The shootout was held in a large conference room at the Pasadena Hilton hotel. As the presentations got under way, Mark was nowhere to be seen. That meant the presentation slides he was putting together were nowhere to be seen, either.

I kept looking at my watch. We were due to present at 1:00 p.m. Around 10:30, still with no sign of Mark and no word from him, I began to get nervous. I called his home. His wife, Diana, answered the phone and sternly told me she would not wake him, that he had worked on a proposal all night and needed sleep. I laughed, but then anxiously pointed out that all his efforts would have been for nothing if he didn't wake up and make an appearance at the Pasadena Hilton very soon.

He made it with thirty minutes to spare, bringing a twenty-seven-page presentation package describing what we called the "Mars Mobile Pathfinder" in order to make sure the "mobile" idea would come through from the first.

When our turn came, we noticed that the geologists on the panel perked up on hearing our concept. A moving rover that could examine Mars rocks on a mineralogical level was just what they had been longing for.

Not long after the day of the presentations, we learned that initial approval was being given to two projects. One was an orbiter; the other was ours. Each would have a status that in NASA is called a "pre-project." JPL would receive preapproval funding for both, with the money to be used for examining the concepts in greater detail, identifying the major challenges, and presenting plans for moving ahead. The orbiter project would be called the Mars Reconnaissance Orbiter. Our project was to be called the Mars Exploratory Rover—"MER" for short.

The head of NASA got in touch soon after to say, "I think we should do *two* rovers. If one fails, we'd have a backup, and we could aim for two different places on Mars."

Eventually the two spacecraft would become known as Spirit and Opportunity. I led the flight system engineering team while also managing the development of the challenging EDL system.

For the next few years, I would be focusing primarily on the twin MER projects. At the same time, I would continue to be an active part of a loose-knit group juggling ideas for landing a much larger Mars rover, something we initially called a "MegaRover" but was soon being called the "Mars Smart Lander." Without Pathfinder and Sojourner, we couldn't have built Spirit and Opportunity, and without those two rovers we could never have built what would become the Mars Science Laboratory and its Curiosity rover.

CHAPTER 4
The Birth of MSL

By late 2000, the groundwork was being laid for what would eventually become the Curiosity rover. A NASA-sponsored science definition team had drafted a set of mission objectives for the Mars Smart Lander, to be launched in 2007. Based on assuming our new entry guidance technique would be used, it called for the elements we had been working toward: a precision landing inside a small landing ellipse, with a rover that could travel to a much wider range of locations than previously possible.

The objectives statement anticipated that the mission would "allow exciting new science to be accomplished on the Martian surface and test new technologies . . . including safe and accurate delivery of large payloads, long-range mobility, subsurface access, and complex sample handling."

"Exciting" wasn't just a description of the science but also an accurate description of the reaction of people at JPL who work on flight missions. What had been the informal bubble teams now morphed into a Mars Smart Lander team, again starting with pre-project status that provided us enough funding to develop basic concepts and plans. The initial budget was about the same as the total for the two MER projects—a bit less than $1 billion.

Most of the people at JPL with experience in rovers and EDL were already hard at work on Spirit and Opportunity, leaving MSL with a small team of fewer than twenty people. Fortunately, those of us on MER were able to spend some time helping. The heavy lifting for the design of MSL would have to wait until after MER launched in 2003.

The team was told that the initial focus should be on developing technologies for a large rover. Some of the early areas that the MSL team dived into included re-creating the throttled engine technology used on the Viking missions in the 1970s, developing ideas for imaging radars to help avoid hazards during landing, and developing new subsonic parachutes. On the rover side, they also focused on re-creating the power-supply design of the 1970s and '80s, based on a plutonium energy source that had been used on the Viking and many of the missions to the outer planets like Voyager and Cassini.

One of the problems we struggled with on Spirit and Opportunity would lead to a solution that would revolutionize our ideas for MSL. The entry, descent, and landing plan for Spirit and Opportunity was simply lifted from the airbag design that had been used for Pathfinder. But the MER rovers were much heavier than the load carried by Pathfinder, and this extra weight was pushing the airbags to their limit. The video footage of our airbag drop tests was painful to watch, with scenes of the fabric being torn to shreds under the excess weight. We needed to find a way to reduce the speed of the landing impact. In particular, we needed to reduce the horizontal component of impact—dragging the airbags across the ground—that caused so much ripping of the airbag fabric.

It was some two years since we had dropped the idea of landing a rover directly on its wheels, accepting the arguments that we would probably not be able to control the touchdown speed of a suspended and perhaps swinging rover with enough precision to prevent it from being dragged along the surface and damaged. Now, though, it became clear we needed to take another look at the problem of how to control the effects of wind just before landing, that could cause the rockets to tilt, in turn causing the lander and rover to travel horizontally across the surface. To grapple with this, Miguel and his guidance-and-control team dove into the complex physics involved and developed mathematical equations for the system.

The analysis led to the idea of solving the control problem by adding three small rockets to the backshell, combined with a new camera for detecting horizontal motion; the rover's computer would analyze input from the camera and fire an appropriate combination of the rockets as needed to cancel any swinging motion or tilting. This would avoid excess horizontal motion across the surface. Miguel and I realized that the same math we had used for Spirit and Opportunity also solved the very problem that had earlier stopped us from pursuing the rover-on-a-rope idea. With the right control, any swinging of the rover could be stopped almost instantly.

We demonstrated this for Dara, and he grasped the significance immediately. We now understood how to control a payload on the end of a rope slung below a descent stage, and we agreed that we had been premature to kill off the rover-on-a-rope back when we originally came up with the idea. The possibility of landing a rover on its wheels was now no longer inconceivable.

You can demonstrate this for yourself. Tie a weight to the end of a string and hold the other end of the string with your fingertips. With your free hand, start the weight swinging. As the weight swings in one direction, move your fingertips in that direction; when the weight reverses swing, move your fingertips in the new direction. Most people are able to learn to stop the swing in very few attempts. Once you've mastered it, try letting someone else start the swinging. Then try with your eyes closed.

That crazy idea to land a rover directly on its wheels from early 2000 suddenly no longer looked crazy.

By this time, in early 2002, Tom Rivellini's pallet lander was presenting all sorts of technical difficulties. He had built scaled-down versions of his ultra-lightweight structure and dropped them on slopes and rocks. To his frustration, the design kept failing. As Tom well knew from his work designing and testing Pathfinder's airbags, mechanical structures designed to come into contact with the Mars surface often provide unhappy surprises when tested with real rocks and slopes. Nature often wins and the structure loses. Tom's handsome design was losing.

One day in late July, I had lunch with the chief technologist of the MSL project, my former mentor, Bob Rasmussen. We shared what we knew about Tom's troubles. I asked him if he had heard about the idea for lowering the rover to the ground directly on its wheels. To my surprise, Bob hadn't heard anything about it. I told him the story and added what Miguel and I had learned about controlling pendulum-like dynamics for Spirit and Opportunity.

Bob was intrigued. Being a former guidance-and-control engineer, he was especially fond of the finesse that such a system might provide compared with the rather brute-force landing approach we had designed for Spirit and Opportunity. He wrote an email back to the then-MSL project manager, Mike Sander, and to the rest of the project saying we should really consider this. The following week I went to Mike's office and gave him a short briefing on the idea. Mike was interested but clearly skeptical.

In late 2002, with launch preparations being made for Spirit and Opportunity, Dara Sabahi and a few of us finally had time to help MSL. Dara quickly rounded up JPL's core EDL gang, about fourteen people including me, and held a series of brainstorming workshops. Our group looked at the suspended-rover idea and evaluated it against other ideas including the pallet design.

Even Tom, who for the past two years had been trying to get the pallet approach to work, quickly began to agree that the suspended-rover approach made sense. Landing the rover on its wheels would solve a lot of technical problems. Tom could see that our rover-on-a-rope idea looked more achievable than his pallet. But it also created new problems and seemingly countless questions. For example, when would you lower the rover? Up high when still attached to the parachute? Or would the descent stage hover and lower the rover just above the ground?

How long would the ropes have to be? When would we release the stowed wheels—before the rover was suspended on the ropes, or after?

Once the rover was suspended, could the rocket plume damage it?

Did we have to do something special with the wheels? Would they need to lock in place?

How would the rover get released from the three ropes that lowered it?

How would the computer controlling the landing detect that the rover was on the ground?

Once landed, how would the ropes be prevented from draping on the rover?

What would happen to the descent stage? Would it fly off and crash in the distance? If so, how far away?

These and many others were still to be answered, but all of us felt that we could find the answers with a year or so of further study. What troubled us most of all was that it looked as if the basic idea of rover-suspended-on-a-rope was going to be a hard sell to management. This is not how Buck Rogers would do it. This is not how the Apollo astronauts did it. This is not how the Viking Landers did it. This is not even how Mars Pathfinder did it. This is simply not how you land on another planet.

If there was going to be a big backlash, we had better be prepared to convince everyone up the ladder that this was the right solution.

Late spring, 2003. Spirit and Opportunity had been finished with not a minute to spare and were now at Cape Canaveral, where the teams were preparing to put them atop the launch vehicles.

If this had been a more typical flight project, all the design and operational questions would have been completely resolved, all the testing finished, and the teams at JPL would have moved off to new projects. Missions with the complexity of Spirit and Opportunity take at least four or more years from concept to launch; this pair had been built in only three years from our first presentation. Although the cruise software had been completed just in time, the team had not yet finished testing and tuning the EDL software or the rover's surface operations software. Fortunately, landing was still seven months away.

That's why, even though the two spacecraft were already at Cape Canaveral, we were still running tests in the test bed at JPL. We typically build at least two spacecraft—a flight version that is treated with kid gloves and only briefly tested before it is launched, and another earthbound unit that gets put through its paces and is used for nearly all testing of the hardware design, software, and systems. This earthbound unit is also used for problem solving and more testing while the flight version is en route to its destination. The one that doesn't fly gets to live its life in the test bed or in what we call "the Mars yard"—an outdoor test area strewn with rocks to simulate the surface of Mars.

I dropped by the test bed one afternoon to see how some electrical tests were going on the grounding of the wiring between two parts of the Spirit rover. Two days earlier, the team in Florida had been verifying the grounding of the wiring between two parts of the Spirit rover. Suddenly, the testing halted. After an inspection, they found that a fuse deep inside the rover had blown. This faced us with an agonizing dilemma: Could we fly with a blown fuse? Could we use the test bed in Pasadena to prove that the rover would be able to land and function properly?

Tensions were running high. Back at JPL, standing in the test bed lab, one of the team that had designed the power systems said, "Rob, with the fuse blown, when we cut Spirit's electrical cable to separate the rover from the lander during EDL, here's what might happen: It might cause a short circuit that would open a relay, disconnecting the rover battery from the rover and leaving the rover without power."

I thought, "This can't be right." Standing in the test bed lab, three of us studied the schematics, all of us knowing that if this were true, we would not be launching. There was no time to take the Spirit apart to fix the fuse or to fix the design to prevent the relay from opening. I started feeling dizzy.

What happened next is something I can't explain, something that still embarrasses me. I'm known as a calm guy, someone who can address any problem, any seeming disaster, without losing my cool.

Not this time. I *screamed.* I threw my pen to the ground and it split open, spilling ink—in the middle of the spotless clean room. Everyone froze, shocked, staring at me like I needed to be put into a straightjacket.

I took a deep breath, gathering my anger but not my wits. I shut the book of schematics and announced, "Okay, everyone, we're going home early. We're closing down." I had hastily concluded in those few seconds that it was not going to be possible to launch Spirit and Opportunity in 2003. We could not make the launch date we had been working toward for so long. After a few minutes of calming down, I marched up and told project management about the situation, and called an emergency meeting to see if we could figure out a way around the problem.

At the Cape, the Spirit spacecraft was close to being fully buttoned up for launch. It didn't take us long to confirm that with a blown fuse the cable cutting event could indeed take the battery off line, leaving the craft without battery power just as it approached the most critical phase of the landing. We dived into a brainstorming session working as if we knew there was a solution—we only had to figure out how to find it.

Then someone in the group recognized that there was enough time for the software to put the battery back on line. That was the key we were looking for.

We could add some lines of code to the software so that every time a pyrotechnic device fired to cut one of the cables, the rover would quickly switch back on the rover's battery power relay.

We all felt huge relief.

In the months leading up to Spirit and Opportunity's launch we had a number of these close calls. Even though we were not done, once Spirit and Opportunity were safely off the ground and on their way, we were able to unwind and laugh a little.

On landing day, about 6:00 p.m., after a light boxed dinner, I climbed the steps and took my seat as EDL lead in the Cruise Mission Support Area, our control room. My task this evening was to watch displays of the radio signals from Spirit—signals that we call "tones"—and translate them on the headset for the team and the world to follow.

My EDL chief engineer, Wayne Lee, would be doing the verbal play-by-play for the outside world. Wayne would listen to me as I marked the events with the help of radio expert Polly Estabrook. This would allow us to share the extraordinary experience of watching radio signals from Mars and describe the steps of Spirit's arrival. As always on these landing days, this would be a realization of hopes for the many engineers and technicians who had worked on the project, and a realization of dreams for many scientists and millions of people who had been following our mission on the web.

In the previous few days, NASA had been making it clear to the media and the public that there was a good chance Spirit and her sister rover might not succeed. While I agreed that it was a bad idea to be overconfident, the depressing drumbeat of failure that followed us was getting me down. And not just me: The control room was filled with tense laughter and weak smiles.

Since Pathfinder, we have had a tradition of playing a song for special occasions. We would do it nearly every day during surface ops as well, just because we were happy. I got to pick the tune for that night and thought it might be good to lighten things up. I found a nice quiet moment early in the evening to put on Bobby McFerrin's "Don't Worry, Be Happy." Within a few minutes, the tune did its job. I could see the tension ease a bit. It certainly helped calm my own nerves after a most nerve-wracking week.

Six and a half years earlier, when we landed Mars Pathfinder, many people felt we had just gotten lucky—everything seemed to work, even though it had been done on the "faster, better, cheaper" model. That grousing didn't make the team feel any less gratified by the accomplishment. But now a new generation of engineers and managers had watched as we made every faltering step. After all this, they knew damn well, you don't succeed just because you are lucky, you *make* your luck . . . but then you cross your fingers and eat the lucky peanuts anyway.

I didn't mind one bit that this wouldn't be the first rover to be landed on Mars. Pathfinder and its Sojourner had done that six years earlier. Spirit, with its untethered roving capability, held the promise of offering scientists the possibility of finding out "what the grass looks like on the other side of the hill."

Sitting in the control room gave me a déjà vu feeling. I had monitored and reported the landing of Mars Pathfinder in 1997 and now I would be doing much the same for the Spirit landing.

Finally Spirit penetrated the Martian atmosphere. "The parachute's been detected," I announced with restrained delight. Polly and I could clearly see the change in the radio signal's frequency as the spacecraft slowed. The heat shield came off, and then a tone told us that the lander with the rover stuffed inside had descended on its twenty-meter-long rope bridle.

"Lander separation has been detected!" I chirped. Another tone told me that the radar was working. It had seen the ground.

"Radar in lock."

A tiny step in the plot I was staring at on our monitors showed the rover might be slowing down even more. The rockets were firing.

I reported, "The RAD has fired!"

Or maybe not . . .

Suddenly . . . nothing. The signal had vanished from our screens. It disappeared just as the rockets were firing. I reluctantly announced, "We don't have a signal at the moment."

Minutes went by. Wayne told the audience, "Please stand by."

Silence. I could hear my heart pounding. After about ten minutes of nothing, I began to curse Mark Adler silently for coming into my office with his proposal that led to Spirit and Opportunity. With all the bad moments I had had on flight projects, this felt like the worst.

Fifteen minutes later. I was becoming convinced that Spirit has been lost. Sweat was running down my brow.

Two minutes passed. Polly grabbed my arm. "Rob, *look!*"

My eyes stumbled over the display. There it was. I now remembered: I had asked to have a small backup antenna installed under the base of the lander, in case the lander got stuck sideways on a rock or went upside down on landing and was unable to right itself. EDL system engineer Jason Willis had added code to the software so it would command a swap to that antenna after seventeen minutes on the surface. Those seventeen minutes had elapsed, and the rover had swapped antennas and revealed itself.

I yelled, "We're on Mars!"

In an instant, we were all once again awash in the customary hugs and high-fives.

Three weeks later to the hour, Opportunity safely landed on Mars, this time without any panic attacks. After receiving hugs from flight system manager Richard Cook and from Miguel San Martin, I could no longer contain my tears. Soon I was explaining what we were watching to Arnold Schwarzenegger and Al Gore, who had come to watch, and once again I got a hug from June Lockhart.

I couldn't remember such a remarkable feeling of thankfulness since I held my healthy and beautiful baby daughter moments after her birth, and then I realized I was thinking about Spirit and Opportunity as if they were my children we had brought to life. That's way over the top, but there it was.

On that joyous Opportunity landing day, I had just enough energy to speak at the post-landed press conference, high-five the EDL team, then head home. I hugged my wife, Dominique, my daughter, Caline, and my mom. Over a glass of nice red wine, we toasted the first images to stream in on CNN and NASA-TV.

Opportunity had hit the jackpot. It had bounced into the center of a small crater surrounded by millions of little BB-sized balls of hematite clearly formed under wet conditions. The two-foot-tall rim of the crater revealed a cornucopia of water-altered minerals sticking out of the surrounding soil.

Between them, Spirit and Opportunity during their lifetimes would put in more than sixteen years exploring different parts of Mars, traveling a combined distance of more than 29 miles (46 km). Each rover provided evidence that Mars indeed once had a warmer and certainly wetter environment, one that might have provided an environment for early life forms to exist.

I wondered: If those two spacecraft could hold so much promise for advancing our knowledge of Mars, what would MSL be able to do?

CHAPTER 5
Looking Ahead to Putting People on Mars

Sometimes a request lands in your lap that makes you feel recognized and honored. In 2004, the White House announced a far-reaching, long-term vision for space exploration that among other things called for a "human and robotic program to explore the solar system and beyond." The concept would include human exploration of Mars.

That broad-scale announcement became personal for me just after I landed in Lihue, Kauai, where my wife, daughter, and I had flown to enjoy our first vacation in years. I received a cell phone call from former JPLer Jennifer Trosper. She had worked on our small team on Mars Pathfinder in the 1990s and had led the project-level system engineering on Mars Exploration Rover with me, on which she and her team pioneered Mars rover operations concepts. By then she had become the most experienced Mars surface operations engineer within NASA.

Jennifer and her husband, who was in the Air Force, had recently moved to the D.C. area so that he could continue pursuing his chosen specialty, as a test pilot. She had landed a job at NASA headquarters as what they call a "detailee," someone from one of the NASA field centers who is loaned to headquarters. Word had come down that NASA was looking for people to lead a series of fifteen "capability" roadmap teams—formed around such topics as in-space transportation, human health and support systems, and scientific sensors and other measuring equipment—to help set long-term directions for the agency's focus in the coming decades. One of the teams was to explore "human planetary landing systems," and Jennifer had been asked if she had contact with anyone who had experience in planetary landings. She knew my involvement with landing systems for Pathfinder, Spirit, Opportunity, and the upcoming MSL, and had called to ask if she could put my name in to head the landing systems team.

I was flattered, answering that I was pleased at the suggestion and would be glad to be considered. Not long after Jennifer's call, I was notified that I had been named to head the Human Planetary Landing Systems panel and given carte blanche to select my team. Just putting together a team sounded like a challenge, and I couldn't even begin to imagine the challenges that lay ahead. Still, through the years I had met a large number of EDL experts from around NASA and beyond. I would begin by trying to round up a group of my old contacts. Because of the high visibility of these workshops, presumably it would not be too hard to get the best to show up.

I started by lining up two co-chairs. The first was Claude Graves, who had been a leader in developing the entry systems for the return of Apollo's astronauts from the moon to Earth. (His death in 2006, only two years after our work ended, was a huge loss to the space program.) As the father of the Space Shuttle's entry guidance, his experience with working out how the astronauts might fly a Mars lander to the chosen landing site would provide special insight for the panel. Claude was also a huge proponent for using large lifting body spacecraft (like a wingless shuttle) as a way to get astronauts to the surface of Mars. I was skeptical of that approach but expected he might be able to talk me into it.

My other choice for co-chair was Harrison Schmitt, known as "Jack." He had walked on and driven around the moon as part of the Apollo 17 crew, the last Apollo moon mission. He held a doctorate in geology from Harvard University and had also served in Congress as a senator from New Mexico. When I met him at a JPL review, I couldn't let him get away. I told him about my plans and asked, "Would you be interested in being a NASA consultant as my co-chair?" He delighted me by readily agreeing. Both of my first choices had said yes, which left me feeling pleased and flattered.

For the rest of the panel, the group I rounded up included some core Space Shuttle EDLers, other Shuttle astronauts, and people who had been building Mars landers including Viking. In addition, I also brought in a few key thinkers from industry and academia, including an old friend, Bobby Braun, whose short hair and youthful look belied his status as a Georgia Tech professor and former NASA chief engineer. He had been instrumental in bringing entry-system aerodynamics engineering back from the edge of near extinction. Later he would become the first chief technologist NASA had had in decades.

Others included Marshall Space Flight Center's Michelle Monk and old hands like NASA Langley's Dick Powell, who had worked on early EDL concepts for Mars and would help make sure that our team's work would be solid. In all, I put together fifty EDL people from around the country. This would be the first time in NASA history that EDL people who built "the little ones" like Pathfinder, Spirit, and Opportunity would be formally putting their heads together with people who built "the big ones"—the spaceships that had carried humans to the moon and back, as well as round-trip to the Space Shuttle.

While our focus was on Mars, our charter was actually to address looking at landing people on *any* planetary body. At the time, many people were saying that they knew how to land humans on Mars. I honestly wanted to know how it could be done. You might think that as the guy who had landed more stuff on Mars than anyone else on the planet, I would know what it would take to land anything there. The truth was that I knew about landing *small* things, things the size of a dining room table. I knew little about landing anything big enough to carry people.

Four months after the first phone call from Jennifer, my full panel gathered in December 2004 for a day-and-a-half session nearby at the California Institute of Technology. This was going to be eye-opening for me. Human missions to Mars would need to land many tons of gear in order to provide for a few astronauts to survive on the planet. In fact, given the struggle my teammates on MSL and I were having to land a one-ton (900 kg) rover, I had trouble imaging how it would ever be possible to put on Mars the kind of supplies that would be required. I had even more trouble just getting my head around the idea of a house-size spacecraft.

It wasn't until everyone had arrived that I realized, looking around the room, how this assembly probably held more knowledge on the subject than existed anywhere else in the world. Someone even ghoulishly joked that a bomb in the room would set back US space efforts twenty years or more.

I called that first session to order feeling both curious about what work had been done over the years and, as well, deeply concerned that the level of engineering needed to define a human-scale Mars EDL architecture might not actually have been invented yet.

To kick off the session, after everyone had introduced himself or herself and provided a brief background, I asked them to share what I called their "care-abouts." I wanted everyone to get their favorite technology or pet peeve on the table right off the bat. I needed to know where people were coming from. What did they think the major issues were?

Kent Joosten, a brilliant human-Mars-mission design architect, explained NASA's "short- and long-stay" mission designs. Long stays would require the astronauts to be on Mars for 545 days (530 Martian days, or "sols") and be away from Earth for two and a half years. Short stays would require "only" forty days on the surface of Mars and only a little more than one and a half years away from home. In either case, the stuff that needed to be landed on Mars for the astronauts would be massive.

From Kent, we heard the description of a whole stocked-up camp that would be placed robotically prior to the astronauts' arrival. To save weight, water would be extracted from the Martian air—though we would later learn of an easier way—and the oxidizer for the fuel needed to take off for the journey back to Earth would be manufactured on Mars using what he called "in-situ resource utilization": Radioactive power plants and solar arrays would be brought in by spacecraft, ready to be set up. Even the crew's Mars ascent vehicle for the return home would be positioned in advance, fueled and ready to go.

Jack Schmitt next took the floor and explained what a Mars mission would look like from an astronaut's point of view. Being in space was lonely enough, he said. Astronauts would need people to talk to, people in Mission Control to help work them through the tough spots. But Mission Control in Houston is between eleven and twenty-two light-minutes away, depending on where Mars and Earth are located in their orbits. You could talk for ten or twenty minutes before the first word you uttered ever arrived at an Earthling's ears. Jack said that there had to be a support team in orbit to help the astronauts as they descended and then as they went about their work on the Mars surface.

Would the astronauts left in orbit who perform Mission Control stay there for the duration of the surface mission? Wouldn't they want to come on down and join their colleagues on the surface?

"I don't know," Jack said. "Maybe they could swap places, each acting as Mission Control for the other."

An unexpected issue surfaced when Jack Schmitt commented, "To do this landing safely, of course we need the ability for the astronauts at any time to hit the abort button, wave off the landing approach, and go back into outer space." He said the Mars EDL designers would need to provide this ability. Bobby, Dick, and I glanced at one another, appalled, and we all looked at Jack with stunned expressions that translated as, "What? Are you kidding? How the heck could we do *that*?"

One of the astronauts piped up: "Well, the Space Shuttle had a supersonic launch-abort mode." They had simulated turning the craft around and firing backward so that it could return to Florida in case of a problem. Fortunately, no Space Shuttle crew had ever needed it. Dick and Bobby introduced the NASA Langley engineers who had designed and analyzed the abort procedure for the Space Shuttle. What they told us wasn't reassuring: The Shuttle's abort would probably never have worked. It was a Hail Mary. In that kind of emergency, they explained, "It may not have worked, but you'd be dead in any case."

With Mars landers, aborting raises a much bigger problem. All of our Mars landers and rovers as they approached the planet shed a Rube Goldberg–worthy assemblage of intricate hardware and components, leaving a trail of debris as the vehicle undresses itself on the way to the surface. We assumed that a human mission would do the same. How could it do all of those transformations and undressings, shedding hardware, while still being able to change its stripes and turn around to fly back up into outer space in an emergency? It would be worse than if the Shuttle designers had been told to make it able to change its mind about landing in Florida and zoom back into space. The thought was daunting.

When we explained that the Mars entry was more like landing on Earth than landing on the moon, the astronauts were quite surprised. We gave them an explanation that went something like this:

"Imagine you're going at Mach 15 surrounded by a bubble of hot gas as you plunge through the Martian upper atmosphere. You're decelerating at up to 6 gees." (That's six times the normal Earth pull of gravity, sometimes written as "6 G's.") "If you wanted to change your mind and head back to space, how would you do that? You're going so fast that there's no way to undress yourself from the heat shield and turn your craft into a rocket. Worse, you're headed in the wrong direction going extremely fast. The amount of additional fuel you would have to have at that point to get back into space would be enormous."

The astronauts on the panel didn't like the idea of not being able to abort the landing. But they understood the reasoning and finally, if reluctantly, accepted the inevitable. Jack suggested an alternative: an "abort to surface" approach. "If there was a problem during EDL and the astronauts were forced to land many miles from the predeployed Mars outpost"—where housing, supplies, and fuel had been landed—"we'd need to be assured that eventually we could get there. So we'd need to have a long-distance driving capability."

At least that was something we could imagine engineering into the lander. I thought it better not to point out that if the crew were so many miles off course, it was unlikely that whatever caused them to be that far afield would mean there would be nothing left to worry about.

The robotic EDLers in the crowd were thrown for another loop when Shuttle Astronaut Commander Barry (Butch) Wilmore strongly articulated the need for the astronauts to be able to take over control during landing on Mars, just as Neil Armstrong had to do during the first lunar landing. I know that if I were an astronaut, I'd want to be able to control my spaceship. But those of us who design rovers to land on Mars also know that those landings are nothing like landing on the moon or landing the Space Shuttle back on Earth. The Martian atmosphere is so close to the surface of the planet that events during landing unfold at lightning speed.

We're talking about an astronaut who has been traveling through deep space in zero gee for well over a year and is now undergoing extreme entry deceleration. Silicon chips running at 120 million operations a second can handle it, but I wouldn't expect any human—not even a highly trained, exceptionally fit astronaut—to be able to do the same.

John B. Charles and NASA Flight Surgeon Jonathan Clark argued that even with exercise, it was doubtful that the crew would be in any shape to make snap judgments during the dramatic landing. We were surprised to learn that even some of the Shuttle landings were less than elegant, with crews making some close calls during their benign, supposedly routine touchdowns. Some overshot the landing, others veered to the left or right.

So, following a twenty-three-month journey in zero gee and the rapid deceleration on arrival, the astronauts might barely be able to move or walk off their lander, let alone take control and "fly" their spacecraft to a landing. The entire sequence of events for entry, descent, and landing would almost certainly need to be automated. Pilots would have to rely on preprogrammed software EDL behaviors, the same as we'd been using for our robotic astronauts. Nevertheless the designers of human landing systems will bend over backward to provide human pilots the ability to make choices—even if there are few to make.

Over the course of three months, we gathered for two additional multiday workshops—the second at the NASA Ames Research Center, in Mountain View, California, and the final one at the Johnson Space Center, in Houston, Texas. Gradually we started to home in on a few key observations and details about putting human-scale landers on Mars. It was a credit to the group and how well they worked together that there were plenty of eye-openers, ideas that were daunting and exciting.

One item, as an example: As a Mars astronaut lander undresses itself, large pieces like the heat shield tumble and speed toward the surface, landing with the force of an explosion. For human missions, this used-up and now useless hardware detritus would be massive. A lander carrying a full load might require a heat shield alone weighing as much as a 727 aircraft. The explosion of an object like that as it hit the surface two or three times faster than the speed of sound would easily be enough to destroy a nearby base camp established in advance.

We would either have to learn how to design EDL systems that did not shed hardware, or we would have to aim the astronaut's spacecraft far from the planned landing site and then, in the final moments, use rockets on the lander to quickly divert it to the designated landing site, while its discarded equipment rained down on Mars a mile or two away.

What became clear to me was that in the fifty years of its existence, NASA had never really had the focus, resources, or real Mars landing experiences to study these problems clearly. Hardly any of the people gathered for my sessions had ever thought about the problem end to end. How exciting it was to consider that if anyone could come up with answers, it would most likely be the people gathered here.

As the conversations gained focus, we began to narrow in on a few clear challenges. One issue that was primary for me is representative of the back-and-forth that the panel went through on many topics: The MSL team had spent the last three years struggling to figure out how to land a one-ton rover on Mars. The pages of items astronauts would need for a longer stay would include food (there's not such good hunting on Mars), fuel, oxygen, energy, breathable air, some type of living quarters, roving equipment, and a return vehicle. And that's not even thinking about how they would get rid of their waste, bathe, and clean their eating utensils and their clothing. And what if there were a medical emergency or an illness needing a prescription not in the medicine chest? Even if we counted on presupply missions, the initial landings would probably require a spacecraft weighing 30 to 70 metric tons (27,000–63,000 kg). How could we possibly bring that much stuff to a stop at a precise location on Mars?

In the group were two of the nation's leading experts in parachute design. Juan Cruz, from the NASA Langley Research Center in Virginia, is a part-time professor at Georgia Tech and has a reputation as NASA's best expert for supersonic parachutes and supersonic deceleration. He and I knew each other well, since he had worked on the chutes for Spirit and Opportunity and was deeply involved in the design analysis for the MSL parachute. Hard to miss in a crowd, he has long, wavy reddish-brown hair cut into a rocker's mullet. (For the record, he has since given up this hairstyle.)

We also had on the panel a longtime builder of Mars parachutes. Al Witkowski, of Pioneer Airspace, is one of the world's most prolific designers of aerospace chutes. His passion for parachute design and manufacturing details is the complete opposite of Juan's erudite focus on the physics of parachutes.

Al is happy when he holds and folds, but he is happiest when one of his parachutes is delivered and installed on yet another Mars spacecraft.

When the conversation turned to decelerating from supersonic speeds, the first item addressed was parachutes. Juan explained the fundamentals and design limitation of large supersonics chutes. He reminded us that the largest ever tested supersonically was about 85 feet (26 m) in diameter.

But the people who had been asked to do human Mars EDL simulations years earlier reminded us that they had done computer studies of simulated but much larger parachutes that would allow a human-scale lander to slow from somewhere just above Mach 3 down to less than Mach 1 over a few tens of seconds, but with a rather large back-breaking jerk as the parachute inflated.

Al and Juan just laughed. Juan told them, "We can't scale parachutes up to the size of the Rose Bowl! Not without understanding the physics of parachute deployment and inflation of something that size. We just can't say, 'All we have to do is make them bigger.' We can't even say that the parachute will open in time before the lander hits the ground."

I asked, "What about *clusters* of supersonic chutes?"

"Maybe, but there's a complex dynamic plus aero-heating that could melt your parachutes before they even begin to slow you down."

I asked Juan, "Are you telling us we don't have any parachute solution for going from supersonic speed down to subsonic speed for a large-scale lander?"

"That's right," he said.

The room went silent.

We created a subgroup to brainstorm various technology options to address the issue of slowing. Its members came back with an alphabet soup of possible technologies. Supersonic retropropulsion, mechanically deployable entry system, hypersonic inflatable aerodynamic decelerator, supersonic inflatable aerodynamic decelerator, even multistaged parachutes. Any of these individually or in some combination might work with the right design, analysis, and testing.

By then it was beginning to sink in: At this stage, none of the experts was able to offer a single effective way of slowing down. We had a list of interesting possibilities, but no one in the room could say which of these, if any, would work. We would need to invent new ways to achieve this essential element. But with current knowledge and experience, not one of us, Juan included, was confident we could even come up with an idea that had a high chance of working. Each of the possibilities had significant problems that would have to be overcome, and we were a long way from that.

We spent the last day brainstorming a schedule or timeline that we would need to solve the EDL problem and get it to the point that it would be usable for landing humans on Mars. We figured that, like the Apollo project, one or more Earth flight tests of bits and pieces of the full-scale EDL design would be required.

Although I was not a big fan of the idea, the team also assumed that we would at some point land at least one smaller (perhaps at one-tenth scale) robotic Mars mission to test some of the new technologies. When you stacked all of these up and counted forward, we calculated that the final EDL design concept would have to be ready by 2015. On that timetable, the first human footprints would not appear on the Martian soil until about 2032, and that would be *if* the project were funded soon.

In these incredibly fascinating workshops spread over six months, my human planetary roadmap team had created guidelines for dealing with some of the more challenging technical problems. Since then, with Bobby Braun's indispensable help, NASA has gradually taken up the call to move forward. As an outcome of our efforts, some of the needed technologies are actually being developed right now.

CHAPTER 6
The Challenges of Landing

The Human Planetary Landing Systems chairmanship was of course a side-bar to my work at JPL. Early in 2004, I was pleased when JPL management tapped Peter Theisinger to be MSL's new project manager. Pete and I went back a long way together, most recently from his role on Spirit and Opportunity, where he had been serving as project manager.

A project manager is the captain of the ship, the person responsible for making sure the project is on course and moving ahead smoothly. He or she is in charge of the schedule, and the allocation of money. But most of all, the project manager is responsible for maintaining relationships with the people providing the funds. On a project as big as MSL, this is a highly challenging job, one that requires a lot of lieutenants. (And not unrecognized: In 2013, both of the people who served as MSL project managers, Pete Theisinger and Richard Cook, were named to *Time*'s list of the "100 Most Influential People" for the year.)

With white hair and white mustache, Pete has an approachable, Captain Kangaroo–like look. When relaxed, his staccato laugh can be infectious, but for the most part he holds it far in reserve. He is very particular, if not obsessive, about correctness and clarity of language and thought. When asked to adjudicate a technical issue, he'll start his commentary with a few words, then stop in midsentence to collect his thoughts, meanwhile holding up his hand like a traffic cop so no one will try to jump in before he has finished. Heaven help those who give a presentation to him with an ill-conceived or illogical argument. Pete will find the holes and pounce. Unfortunately, many engineers are poor communicators. I have sometimes watched in horror as someone's solid presentation went down in flames because of poor communication style despite excellent technical content.

Yet Pete never comes across as a salesman for his projects, even to the point of reassuring management, "We're going to cancel this project unless we can do it right." The attitude works well for him, making him believable and powerful. I learned that approach from him, and it has worked well for me.

Once Pete had settled into his new job and had a chance to grasp the basic concepts we were pursuing for MSL, he became concerned about the entry, descent, and landing design, and wanted consensus for this seemingly outrageous idea from people outside of JPL. He gave the assignment to Jeff Umland, a PhD who was our EDL design lead at the time.

Jeff got some help in rounding up landing experts from around the United States, assembling a group that included a top Apollo landing engineer, an Apollo astronaut, a Mars Viking lander guidance-and-control engineer, a McDonnell Douglas controls engineer, and a Sikorsky helicopter pilot. I worried that if the EDL gang couldn't convince these folks we were on the right track, we would be stuck with the traditional architecture: a three-legged lander with a massive rover sitting on its roof. I had been down that road, and it was an approach I didn't relish struggling with.

For all the high-powered expertise, this wasn't even an "official" review board. Instead it was a toe in the water to see if our concept would pass what we nervously thought of as a "laugh test": Would they think the design so outlandish that they would laugh us out of the room?

They didn't. So Pete arranged for a follow-up, a formal review by a panel that would have the clout to permanently bury our design. We were given four months to flesh out the details.

The second panel included experts in mission assurance, propulsion, radar, reliability, systems, guidance, multibody dynamics, and kinematics, one of the astronauts from the original panel, and a helicopter pilot. Our presentations were so detailed that they included a propulsion system plumbing diagram, data on the strength of the wheels to handle touchdown, and our plans for testing. The reviewers had a chance to scrutinize our thinking and open up about what they didn't like and what their concerns were. Most of the people on the review panel agreed that, "You should try it, but there's still a lot of the design that you haven't explored yet."

One key sticking point was the touchdown detection. Many thought not including a touchdown sensor seemed crazy, and despite Miguel's excellent discussion of how he would tackle the risks of the rover swinging below the descent stage like a pendulum, they were still worried. They said, "What about excessive winds? What if the rover slid down a slope as it touched the surface? Why don't you have a hazard sensor?" And there were a truckload of other concerns, as well. While not antagonistic, the responses weren't at all what we had been hoping for, but the panel members were positive enough that they urged us to continue.

The pilot pointed out that experienced heavy-lift helicopter pilots can control both the speed and the position of their suspended loads with exquisite precision. This was a man who had extensive experience in one of the early heavy-lift copters, the Sikorsky sky crane. Afterward we started to call our landing approach the "sky crane maneuver."

In November 2004, I agreed to take on for the second time the job of chief engineer for the Mars program. Unlike a project chief engineer, a program chief engineer covers all projects in the program. In this case, that included not just MSL but also projects already at Mars (Spirit and Opportunity rovers, Mars Global Surveyor, and Odyssey orbiter), and projects in the process of being designed, built, and launched (Mars Reconnaissance Orbiter, Phoenix, and Mars Science Laboratory).

The Mars program chief engineer works with the chief engineer for each of the individual Mars projects to make sure common problems or problems that cut across more than one project get solved. These need to be coordinated so that, for example, orbiter missions can support the communications needs and landing-site selection for lander missions. The landers should be able to support the orbiter missions by taking "ground truth" measurements that can help explain what the orbiters are seeing from so high above the surface. In some cases, technical problems on one Mars project would be relevant to another, since they often share similar technology such as computer memories. They would also help each other by keeping an eye on the Mars environment. For example, Mars global dust storms can be detected and the information passed to the rover teams, which then have time to prepare.

By this stage, Spirit and Opportunity had provided evidence confirming that Mars had been warm and wet. NASA headquarters and its Mars Program office decided to redefine MSL's mission. Rather than emphasize the project's technical goals, they were now calling for something far more elegant and significant: advancing humankind's knowledge of the science of Mars. The "Mars Smart Lander" was rechristened "Mars Science Laboratory," conveniently retaining the same initials. Spirit and Opportunity had not been able to show that Mars was once habitable for organisms like those on Earth. Perhaps MSL could.

But how big should this new rover be? The first answer was, "Since we don't yet know how much room the instruments will require, you need to make it as big as possible." On the question of weight, we had two issues: The weight of the laboratory instruments, which for some time would remain a big unknown, and the quite considerable weight of the "sample-handling" equipment—the devices that would gather a piece of soil or rock, filter it as needed, bring it to the instrument compartment on the rover, and distribute the material to the appropriate laboratory instruments.

We guessed at a weight of about 200–300 pounds (90–135 kg) for the instruments. To carry that size and mass of equipment would require a rover about five times heavier than any earlier Mars lander. If that is what we needed to do, we now had in front of us the biggest challenge of our careers: Not just for conversation but in the real world, how do you create a lander *five times bigger* than ever done before and put it safely, gently down on the Martian surface?

The project had not yet moved out of the initial preapproval status, and we had a dagger hanging over us. Instead of solving the many real technical issues, we spent much of 2003 through 2005 costing, recosting, and rerecosting, designing and redesigning—desperately trying to figure out how to keep costs within a range likely to be approved, without lowering the likelihood of success or cutting back on the science by assuming fewer instruments (the "payload"), less weight for the instruments, less space for the instruments, or the like.

Anyone could have gotten whiplash watching the flip-flop of our design changes. The tiny MSL team was so busy trying to make it cheaper that there was not a lot of moving forward. We were torn between directives from NASA headquarters to make it reliable, from the scientists to make it capable, and from JPL's project management and NASA headquarters watching the dollars to make it less costly. The team would take six months settling on a design, produce a detailed cost estimate, and then realize that it was still more expensive than NASA headquarters would approve. They would have to go back and try again.

Aside from the instrument payload, other options considered in this design/cost estimate/try-again cycle included:

- One small supersonic parachute, followed seconds later by a second larger subsonic parachute instead of a single gigantic supersonic one
- One instead of two power generators
- Two identical smaller nonredundant rovers instead of one bigger rover with "full redundancy" (meaning backup units for all critical items)
- Two robotic arms instead of one
- One large antenna to Earth instead of a small one for Earth and another small one for relaying via an orbiter

When we had first conceived of the descent stage, I had counted all along on having at least two computers: one on the rover, and a separate one on the descent stage to handle the massive, critical EDL tasks, as well as handling communications, firing the change-of-course thrusters, and all the other tasks during the seven-month journey. This split made a lot of sense to me. I could easily envision our landing configuration becoming a generic landing system, lowering any kind of payload to the surface of Mars hanging from ropes. We could design and test the descent stage with few cares about the details of the thing that would hang below.

Despite my personal pleading, NASA headquarters and the JPL project management cost analysts decided that providing a full-up computer for the descent stage was too expensive. Instead, we had to design a rover computer that could do double duty. Unhappily, this decision not only increased the complexity of the rover but also threatened its reliability. It also meant that the EDL test team had to fight intense battles with the rover teams for access to the test bed computers, and with the software team to test their software. This frequently turned the test bed into a war zone, with frustration and anxiety flaring between the rover test team and the cruise and EDL test teams.

These design flip-flops continued beyond MSL's first mission concept review in October 2003, and beyond the second over a year later. Only by late 2005 did we finally reach the point of having a solid system design concept. On its journey to Mars, the rover, like a Russian doll, would be tucked up within the aeroshell, made up of the backshell and the heat shield. On top of the aeroshell is the cruise stage, which enables the rover's journey through space by providing it with electrical power, sun- and star-sensing equipment, and radio antenna, as well as housing the thrusters and fuel for making course corrections along the way.

The backshell provides thermal protection for the rover and descent stage during entry, as well as storage for the parachute. The descent stage contains

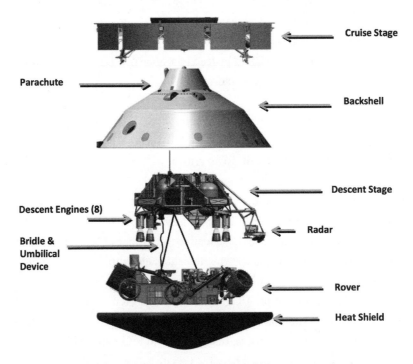

FIGURE 8. Major components of MSL. During the flight to Mars, the descent stage and Curiosity rover are enclosed within the "aeroshell," comprising the backshell and heatshield. (Courtesy of NASA and JPL/Caltech)

the bulk of the hardware needed for landing: the thrusters used during entry and the large throttled engines used for final descent and landing, as well as additional radio antennas. The "bridle and umbilical device" is the equipment that lowers the suspended rover and pays out the nylon ropes and electrical cable that attach the rover to the descent stage during the sky crane phase of the landing.

Creating something that has never been done before often means coming up with ideas that most people, even experts in the field, might find hard to swallow: new ways of seeing that lead to out-of-the-box conclusions, sometimes defying our preconceptions. Who at the time, after all, would ever have believed that a pair of bicycle repairmen would be the first to create a flying machine that actually worked?

That's the situation we were in with our concepts for the entry, descent, and landing of MSL. We had come up with solutions that we were confident about, but they were novel, untried, and for the most part unprovable until we could see that they worked in the Martian atmosphere. Building Mars landers, we are, as mentioned earlier, always haunted by not being able to test EDL because the environment on Mars is so different. On Earth, we cannot reproduce the high-drag forces and heating of entry, we can't test the complicated dynamics of parachute deployment at high speeds, or the lower gravity of Mars, which is only three-eighths of Earth's. Even in a test chamber with the pressure reduced to simulate the atmosphere on Mars, because so many other factors are wrong or missing, the results might be completely misleading.

This was true for all previous Mars missions, and likely it will be true for all future ones. We have to break up EDL into constituent pieces, testing each individually, and for elements that can't be tested on Earth, we have to rely on computer simulations—massive efforts that involve tens of thousands of parameters, along with an extraordinary amount of work to confirm the accuracy of the results. Testing each piece separately gave us confidence that each individually had a high probability of success. With large parts of the system, we understood them well, and I had extraordinarily high confidence in those areas. But there were too many areas where we didn't have good models of how the system worked, or where a simulation could not help.

For example, the electrical subtleties of the sky crane maneuver. Although we could simulate and test the mechanical dynamics of most of the pieces, we could not effectively test the effects of static electricity generated when the rockets fire or all of the electrical currents that flow when the pyrotechnics are triggered. Any electrical interference could disrupt operation of the rover's computer at the critical moments of landing, and even if the rover landed safely, the descent stage might not receive its command to fly away, leaving it to land on top of the rover. In lieu of testing, we had to sharpen our pencils by studying the design to try to convince ourselves that bad things would not happen. To prevent excess static buildup, we added sharp needles to the rover that would bleed off any charge. To deal with excess current from the pyrotechnics, we designed and redesigned the circuits so that they could handle the extra current.

We can't say, "If we land this twenty times, we're confident our equipment will work nineteen of those times." Our maiden voyage, our first flight, is our last flight. Imagine if for its first test flight every new passenger jet airliner had to take off loaded with passengers. That's pretty much what it feels like for us when we face a landing on Mars with these complex, never-completely-tested landing systems.

One of the biggest challenges we faced had been how the EDL software could determine when the rover had settled onto the Martian surface. On some of the prior legged-lander Mars missions, each of the three legs had a sensor that signaled the lander computer that it had touched down. As soon as two of the legs were down, the computer would turn off the engines. This was a crucial element; if the engines were not turned off quickly enough, the rockets could force the lander to bounce and roll over.

MSL was neither a legged lander nor an airbag lander. We had one big advantage over the legged lander. Because our rocket thrusters got no closer to the surface than about 23 feet (7 m), we could afford to land slowly without worrying about digging holes in the ground with the force of the plumes from the landing thrusters. But how would the computer know when our rover had landed? Would we need to add complex touchdown sensors in each of the rover's wheels, as had been suggested much earlier?

It was Miguel San Martin who stumbled on the answer. Mig, always curious, had been playing around with some of his computer simulations. He noticed that the throttle setting dropped by half when the rover made contact with the ground. He realized that the rover weighs about as much as the descent stage does at this point—so an instant after touchdown, the thrusters would have only half as much lifting to do. At first he didn't get the significance, but quickly realized this was an "Aha!" moment. He had stumbled onto the solution.

Imagine a helicopter pilot lowering a heavy container to the ground. In flight, he needs to keep a lot of power to the main rotor in order to counteract the downward pull of the container. As he descends gradually toward the delivery spot, he's monitoring carefully; he knows that as soon as the container is on the ground, this extra power will cause the helicopter to start rising, so as soon as it begins rising, he will quickly roll off the throttle. The start upward is all the confirmation he needs that his container had landed.

What Miguel realized was that the moment of needing less thrust was the moment that the rover had touched down.

Once Miguel's trick had been pointed out to us, we could hardly believe it. This issue we had been struggling with for months had a quite simple solution that required no special touchdown detection sensors at all! The rover software had to do just two things for touchdown: maintain the slow and steady descent rate during the sky crane phase, continuing that slow rate until the ropes start to become slack, and then monitor the engine throttle setting to be sure that once the power has been cut, the situation remained stable, confirming that the rover is firmly on the ground.

Once landed, it would have transformed itself from an interplanetary spacecraft to a scientific remote-controlled rover in less than seven minutes.

CHAPTER 7
The Right Kind of Crazy

There were a lot of things that could go wrong during entry, descent, and landing. It's a long chain of events—thousands and thousands of them.

All the sequences, all the deployments, all the calculations and decision-making that the computer has to do, all the separations: Each one is a link in the chain. Each of these thousands of events requires a lot of things to work the way they are supposed to. Only if every single link of the chain works perfectly do you actually land safely on the surface. If a single one doesn't work, it's game over.

We were able to build some redundant links in the chain of events, but there were just so many places where that wasn't possible. You have one parachute, one structure. All eight of the entry thrusters and all eight of the big throttled rocket engines must work. There are seventy-nine pyrotechnic devices that have to go off at exactly the right time, in exactly the right order.

Mars Science Lab was the most complicated lander we had ever built in terms of EDL complexity. Ironically, it was at the same time the safest and most robust, but also the easiest for something to go wrong—safest in the sense that, for example, it was designed to correct for wind conditions, and able to handle slopes, rocks, and most any conditions that Mars might throw at it. This was a highly reliable design. On all of the earlier Mars missions, whether we were using airbags or legged landers, we well knew that there were places the spacecraft might come down that offered rocks and slopes it could not avoid, or wind gusts or dust storms it might encounter, that could cause the vehicle to be destroyed on landing. We had been simply counting on those conditions not being there at the time of landing.

On the flip side, the previous landing systems were a lot simpler. This meant that the probability we had made an undetected design error was much lower. With Pathfinder, for example, I knew that there was a one in ten or one in twenty chance the landing wouldn't work because of the kind of situations that Mars might present, even something as simple as a wind gust at the last moment. We had built in protections against all those kinds of problems on MSL, but because of the complexity, we couldn't avoid a certain degree of risk that somewhere we had made a design error. In the end, this is simply something we have to live with.

In 2005, NASA Administrator Mike Griffin came out to California for one of his first visits as NASA's top boss. Several of us were proudly showing off the Spirit and Opportunity operations area, when he unexpectedly said, "I need you to decide now: 2009 or 2011." The dates didn't need any explanation: Could we be ready to launch MSL the next time Mars was in close proximity to Earth, three years away? Or would we need two extra years for the subsequent launch window?

We had only recently designed, built, launched and landed both Spirit and Opportunity in just three years. Looking back now at our frame of mind when Mike asked his question, I think that achievement had simply made us overconfident. We were still patting ourselves on the back. Collectively, we were carried away by self-confidence.

If I could relive that day, I would say, "We almost never get the design right on the first pass. We design a piece of the hardware, build it, test it, find out what's not working the way it needs to, have it fixed, then test it again before integrating it into the spacecraft." In retrospect, the 2009 launch date simply did not give us enough time to design, build, try out, and rebuild the hundreds of thousands of components MSL would need. Even so, we all confidently said we could be ready for a 2009 launch.

It didn't come as a surprise to find out that there were folks at NASA headquarters who thought our EDL design was proof that some of us at JPL were losing our minds. There was enough grumbling and frowning that we got word from headquarters only months after the 2009 launch date had been set that Mike Griffin wanted to hear, and wanted his top lieutenants to hear, the MSL story firsthand, in order to pass judgment on "this sky crane thing."

I was part of the JPL team traveling to Washington in my role as Mars program chief engineer, but also because Mike and I knew each other from when he had been on our review board on Mars Pathfinder. He had been open about his admiration for the achievements of our teams on that project. More recently, the reputation of my teammates and me had soared with him after the success of Spirit and Opportunity—I had been on three winners in a row. He had seen how JPL teams and managers operate, intensely technical, dynamic, but not too political, and doing our best to keep everybody up to date about progress on the key issues.

The respect went both ways. A physicist and aerospace engineer, Mike had long ranked high in my esteem as a man of insight and intellect, which is not surprising, since he has more master's degrees than anyone I've ever known, plus a doctorate, plus a spot on that *Time* list of the hundred most influential people.

The JPL group that flew to Washington included my boss, Mars program manager Fuk (pronounced "Fook") Li, as well as the new MSL project manager Richard Cook and the head of the critical EDL team, Adam Steltzner. The pack of us from JPL were doing the coat-and-tie dance, knowing that all the headquarters folks in the session would be dressed that way. When in Rome . . .

We were ushered into a small amphitheater, where we set up our presentation. The audience was going to be not just the top NASA management from headquarters; our dog-and-pony show was going to be videoconferenced to the NASA centers.

The NASA headquarters managers were already long convinced they wanted a big in-situ science rover, but were still unconvinced about whether we could really figure out how to get it to Mars safely. They had heard rumors about the sky crane design but they had not seen details. This concept looked *nothing* like a lander should. Where were the landing platform and its propulsion? Where were the legs?

No one among us had any illusions about how most of these NASA headquarters people viewed us. We can be very confident. To even suggest that you can deliver a rover to another planet takes a lot of arrogance, right? But there is no way we could survive and succeed on arrogance alone. Without a huge dose of humility, we would never achieve anything big.

There is a lot to be said about authentic experience and the school of hard knocks. Yet at the same time, we're sometimes perceived as the kids who wouldn't draw within the lines. It's not that we hadn't known how to follow instructions, but more often that we didn't agree with the instructions in the first place. We were also saddled with a (probably deserved) reputation for going over budget, and with a handiwork record of both dizzying success and embarrassing failure—even though in a business like ours, that should be expected. And now here we were, marching into a major headquarters presentation trying to look and feel confident, prepared to tell the brilliant top managers of NASA that whatever they thought they knew about landing on Mars was passé.

Everybody who has ever presented at a major review knows you need to exude confidence, backed up with hard analysis and, in our case, lots of technical-looking pictures and animations. But you have to be careful. The information presented has to be as accurate as you can make it, but it also needs to be clear and readily grasped. If a presentation is too loaded with technical jargon, the reviewers will wonder if it's just hype.

As someone who has worked with Mike Griffin and seen him in action, I had reminded the others that we should not dumb down anything. Mike liked strong technical arguments based on first principles. It had to be right. If your grasp of aerospace technology wasn't rock solid, he would likely tear you up; he was notorious for it. His deputies, on the other hand, were coming into the room extraordinarily skeptical about our approach.

Admittedly, we at JPL have a tendency to overdesign, so they would be on the alert for that. Also, I knew they were suspicious that we were going to ask for more money—funds that they were counting on using for projects in their own areas of focus. I hoped that if we could get them to understand the challenges, they would be more disposed to be accepting and supportive.

I knew one thing for certain: if Mike liked it, they would have no choice but to go along. The big issue would be convincing Mike of our novel EDL design.

Though I wasn't exactly nervous, I couldn't escape the certainty that if our presentation wasn't totally convincing, the entire project might be shut down, defunded, killed, and we all would go home defeated, wishing JPL had some other project we could jump right into but knowing that there wasn't any such prospect. All we had to do was convince a room full of skeptics that we had a workable solution, even though we wouldn't be able to prove it even to ourselves until our bird had landed safely on Mars.

Mike opened with a quick discussion on a point I knew would be a stumbling block. Given the risks of using parachutes, he felt they weren't worth their weight, that rockets to slow the descent should be all we would need. Adam and I made the case for sticking with the use of our traditional supersonic parachute design. We pointed out that rockets would require more than 1,000 pounds of additional fuel. This would not only put MSL above the launch vehicle's lift capability, but we would also need to develop a whole new technology of flying our rocket engines backward supersonically against the flow, a technique that few had studied much less developed as a reliable way to land on Mars. (This technology, called "supersonic retropulsive decelerators," is now being developed by NASA and by Elon Musk's Space-X team.)

Mike understood and quickly gave up his objections to the parachute. There are few things more satisfying than talking to a person who listens carefully to reasoning, weighs what you said, and is willing to change his mind on the spot if you've succeeded in convincing him.

We were off to a good start. Now it was time for the headliner, Adam Steltzner, a mechanical engineer with a PhD in engineering mechanics who landed at JPL as a specialist in loads and dynamics. He is a colorful storyteller and a captivating speaker in addition to being a first-class engineer and team leader. On his shoulders that day in Washington was the burden of explaining and justifying the sky crane maneuver and convincing the audience, and especially the skeptical Mike Griffin, that it would work.

We had created a video of the landing sequence depicting its rockets, parachute, and its lanyards to lower the rover onto the Martian surface. After we showed the video, when the lights came back on, many of the NASA top brass had turned white: It just looked way too complex. Everybody in the room knew how you were supposed to land, and this didn't look anything like the familiar. It didn't help that everyone in the room knew less than 50 percent of Mars missions had been successful. This audience didn't need to be reminded.

Adam dived into his presentation, a document of thirty-two pages. When Adam reviewed how the rover would be suspended from a single point in the exact middle of the descent stage, Mike responded with a comment cryptic to an outsider but supportive of us: "Oh, yeah—so the rover touchdown dynamics don't couple into the descent stage. Makes sense." That it made sense to Mike was a credit to his mastery of the physics involved. I'm sure some of the people in the room didn't get it.

One part that Mike seemed nervous about was lowering the lander on ropes. Adam went through the design in detail, and Mike peppered him with a series of tough, probing questions. He wanted to know what happens when it touches down. What about pendulum dynamics? What happens if you land on a slope?

We had answers for every challenge. Because of his technical background, he could see we had done the math and had the answers. With interruptions by Mike whenever he had another question or wanted something explained more fully, it took Adam the better part of an hour. When he was done, a lot of the NASA headquarters people voiced serious skepticism. Mike, though, liked what he had seen and heard. He had no trouble with most of what we described.

"This is crazy," Mike announced. "It's the right kind of crazy. So crazy it might just work."

To us he said, "If anyone can pull this off, you guys are the guys to do it." To his own staff he added, "This is the most experienced group of people on Earth for landing on Mars. I think we should give them the chance to make it work."

Adam was trying to be as nonchalant as possible but he could not hide his feeling of triumph. Richard was smiling. Fuk Li was beaming. I was wondering if we should be careful what we wish for.

This turned out to be *the* meeting that set MSL as a "Go" project, overruling a lot of people who had been looking to cancel us and forget the whole thing.

We had gone in knowing we needed a positive statement from Mike to overcome the challenges from his subordinates. Now it was Mike's project as well as all of NASA's. What's more, despite many who still had private reservations, all of the NASA lieutenants now had their marching orders.

It was late in the day as Fuk, Richard, Adam, and I ecstatically walked out of the headquarters building, awash in relief and satisfaction. Just as we jaywalked our way across E Street to catch a cab for the airport and our flight home, a small sports car rushed out of the NASA headquarters underground parking garage. We all leaped aside just in time to see Mike drive by with a look of surprise and a quick wave. His catching us jaywalking was a reminder that we are not that good at staying between the lines.

Mike Griffin's blessings didn't remove all roadblocks. As our budget grew, we would continue to face strong opposition from certain people at headquarters, people with conflicting priorities. We would have to prove ourselves every step of the way.

Not until much later would I recognize a flaw in our progress. We were spending so much time developing EDL and the sky crane, and working to get the blessings of review panels, that the rest of the project wasn't really getting enough scrutiny. We would end up suffering from this shortsightedness.

CHAPTER 8
Scientists at Work

For a number of the scientists whose focus is on other planets, 2006 was a year that brought happy news. The groundwork had been laid in 2004, when NASA headquarters issued an "Announcement of Opportunity." This seventy-five-page document called for proposals to develop instruments that would be carried to Mars aboard the Mars Science Laboratory.

The announcement wasn't any surprise to the instrument, science, and university communities. There had been a buzz going around about the project for many months; everyone in those communities with any interest in research elsewhere in the universe had been anticipating the announcement. Many had already begun putting ideas together.

The announcement calling for proposals began with a statement describing one of the key goals for the overall Mars program. It read, in part:

The overarching goal of the program is to answer the question, "Did life ever exist on Mars?" The scientific objectives established by the program to address this goal are to search for evidence of past or present life, to characterize the climate and volatile history of Mars, to understand the surface and subsurface geology, and to characterize the Martian environment quantitatively in preparation for human exploration.

One common thread that links these objectives is to explore the role of water in all of its states within the "Mars system," from the top of the atmosphere to the interior.

MSL's roles in this larger Mars program called for the rover to:
- Assess the biological potential of at least one target environment identified prior to MSL or discovered by MSL.
- Characterize the geology of the landing region at all appropriate spatial scales (i.e., ranging from micrometers to meters).
- Investigate planetary processes of relevance to past habitability, including the role of water.
- Characterize the broad-spectrum of the surface radiation environment.

What specific types of scientific goals and instrumentation did NASA and its advisory panels of scientists anticipate for the mission? The announcement offered this guidance:
- Analyze Martian atmosphere (gas) samples and/or regolith [dust and soil], rock, ice samples
- Remote sensing investigations with instruments to be mounted on the mast of the rover
- Contact instrument investigations that provide and use instruments to be mounted on a robotic arm
- Investigations that provide and use individual instruments . . . including a sensor to assess the radiation environment at the local Martian surface.

For the companies, individuals, and institutions interested in taking on the rigorous task of submitting a proposal, the requirements even specified the maximum length that would be acceptable: fifty-seven pages, not including the appendixes. The announcement was released on April 14, 2004, with responses due in ninety days. By the deadline date, NASA headquarters had received more than thirty proposals. In the following half-year, the documents were analyzed by an independent review panel. Final decisions were in the hands of NASA headquarters upper management along with NASA's

Mars program director, Doug McCuistion, and Mars program scientist Mike Meyer.

The selected proposals were announced at a time when most people were picking a Christmas tree and buying presents for family and friends. The MSL gang, though, was gathered, listening captivated as MSL payload manager Jeff Simmonds announced the news.

We had all along been geared up for a payload of five or six, or at the most seven, scientific instruments and cameras. I was stunned to hear that NASA headquarters had selected *ten*. Taken together, they would be heavier and would require more space and more power than we had been assuming. My first reaction on hearing the number of instruments selected was shock and dismay. How were we possibly going to accommodate so many?

But as Jeff walked us through each of the instruments and what they were intended to accomplish, I began to understand the amazing discoveries these fantastic tools might lead to and my fears turned into enthusiasm. The advances MSL could bring to humankind's knowledge of Mars generated a level of excitement that was the kind of medicine we needed.

Yet, as soon as we walked out of the room, I wondered, "What was I thinking? This is going to be the coolest thing we have ever sent to Mars but it's going to be damned near impossible to get off of the ground." I was pulled into the excitement of the mission but simultaneously felt a strong pang of anxiety.

I would later learn that the number was higher than we expected because four of the instruments, while offering important science, were more or less freebies. Russia was picking up the cost of one item. A free high-gain antenna actuator would be provided from Spain; the development costs for another instrument would be covered by the side of NASA focused on human exploration of space. A camera was a gift that Mike Malin offered in his proposal, an item already developed for the Phoenix mission but dropped by Phoenix and in effect just lying around with nowhere to go. Each of these represented a valuable addition to the capabilities and science prospects for MSL, at very little extra cost to NASA.

Here are the instruments that survived the selection process, with the principal investigator (PI) of each:

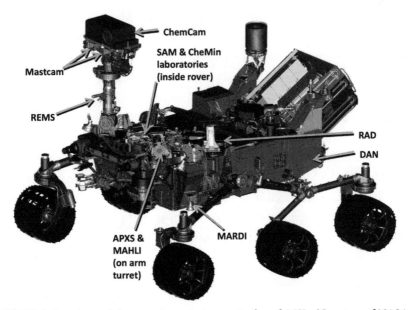

FIGURE 9. Locations of the ten science instruments aboard MSL. (Courtesy of NASA and JPL/Caltech)

THE LABORATORIES

Sample Analysis at Mars (SAM)

PI: Paul Mahaffy, NASA Goddard Space Flight Center

The rock star of MSL's payload, a fantastically intricate "suite" of three instruments: Gas Chromatograph (GC), Quadruple Mass Spectrometer (QMS), and Tunable Laser Spectrometer (TLS). In the final design, SAM would receive rock and soil samples via one of two small doors on the top deck of the rover. It would perform chemical and atmospheric analyses to detect and quantify how much of what chemicals were in each sample, and in the air, by measuring atomic and molecular weights of the material as it is heated and turned into a gas.

GC and QMS work together to first separate (GC) and then detect (QMS) parts per billion levels of complex organic compounds and other molecules. The TLS would precisely measure isotope ratios of carbon and oxygen in carbon dioxide and measures trace levels of methane in the Mars atmosphere. Mahaffy's team at Goddard would design and build the GC and the QMS, while the TLS instrument would be designed and built by a team at JPL.

If Mars harbors the remains of ancient organic life, SAM would be the instrument to reveal it.

Chemical/Mineralogical X-Ray Diffraction Instrument (CheMin)
PI: David Blake, NASA Ames Research Center
CheMin was an inventive new device to look at crystal diffraction patterns revealing the minerals in soil or ground-up samples of rock. Understanding the minerals on Mars can give clues to the environmental conditions that existed when they were formed including the role of water in their formation, as well as providing clues on the origin of the rock, soil and dust.

THE IN-SITU INSTRUMENTS
Alpha-Particle-X-ray-Spectrometer (APXS)
PI: Ralf Gellert, Max-Planck-Institute for Chemistry, Mainz, Germany
Funded and built by the Canadian Space Agency, versions of this instrument had flown on all three of our previous rovers. On the MSL rover, it would be attached to the end of the robotic arm; when placed in near contact with rock and soil targets, it would expose them to alpha particles and X-rays, to provide data on the fraction of atomic elements in the rock targets.

Mars Hand Lens Imager (MAHLI)
PI: Kenneth S. Edgett, Malin Space Science Systems
A camera to be mounted on the turret at the end of the robotic arm, MAHLI would acquire close-up color images of rocks and surface materials in unprecedented detail. (See further discussion in the following chapter.)

THE ENVIRONMENTAL AND SUBSURFACE INSTRUMENTS
Radiation Assessment Detector (RAD)
PI: Donald Hassler, Southwest Research Institute, Boulder, Colorado
RAD was designed to provide data on the radiation levels on the journey to Mars and on planet's surface, toward the eventual landing of astronauts and scientists on the planet. On Earth we are protected from solar and galactic radiation by our planet's magnetic field and thick atmosphere but Mars does not provide the same protection. If we ever put humans on Mars, we need to

know what protection they will need from the solar and galactic radiation that bombards the planet.

Dynamic Albedo of Neutrons (DAN)
PI: Igor Mitrofanov, Space Research Institute (IKI), Russia
DAN can detect water or ice content at extremely low levels (0.1 percent) in layers as deep as a meter beneath the surface. The instrument is designed to send pulses of neutrons straight down into the surface, and by monitoring the delay of the scatter of reflected neutrons (the "dynamic albedo") that come back up from underground, DAN would be able to infer the presence of hydrogen (the "H" in H2O) under the surface.

Like a divining rod, the results could indicate, for example, solid ice three feet below the surface.

Rover Environmental Monitoring Station (REMS)
PI: Javier Gomez-Elvira, Centro de Astrobiología, Spain
The Centro de Astrobiología (CAB) is a joint center of Consejo Superior de Investigaciones Cientificas and the Instituto Nacional de Tecnica Aeroespacial (CSIC-INTA). The instrument, contributed by the Spanish government, is designed to measure items including temperature, pressure, wind speed and direction, humidity, atmospheric dust, as well as local fluctuations in magnetic field. This would become the first weather station on Mars since Mars Pathfinder in 1997.

THE REMOTE SENSING INSTRUMENTS
Mast Camera (MastCam)
PI: Michael Malin, founder, president, and chief scientist of Malin Space Science Systems (MSSS), San Diego
MastCam was conceived as a two-camera system for taking stereo color images and high-definition video at ten frames per second of the terrain to aid the study of the landscape, frost, and weather phenomena, as well as providing images to aid the driving and sampling operations of the rover. One camera has a long-range lens for images up to 0.6 mile (1 km) away; the second is for images as close as 6 feet (2 m).

Numerous instruments from Mike Malin have flown on spacecraft missions since 1986.

Chemistry & Camera (ChemCam)
PI: Roger Wiens, Los Alamos National Laboratory

ChemCam was designed to analyze rocks and soil to determine their compositions, using a laser that fires pulses at a rock or other specimens of interest, and a spectrometer to analyze the results of the laser pulses. The results will help scientists select samples worthy of analysis by other instruments on board Curiosity. (See further discussion in the following chapter.)

Mars Descent Imager (MARDI)
PI: Michael Malin, Malin Space Science Systems

A fixed-focus color camera, MARDI was primarily designed to film the landing, from heat shield separation to a few seconds after touchdown, a period of about two minutes.

CHAPTER 9
The Challenges of Instrument Creation

As a way of providing insight into the process and frustrations of dreaming up and creating an instrument or camera for Mars Science Lab, and having it accepted, here are the stories of two of the people who led the creation of scientific hardware items that flew to Mars aboard Curiosity.

Physicist Roger Wiens, a senior scientist at the Los Alamos National Laboratory in New Mexico, led the team proposing the Chemistry Camera. His interest in Mars was sparked by the Mariner 9 mission, the first spacecraft to orbit Mars. He says that he and his brother saved up their paper-route money to buy parts to build a telescope, which they mounted on a fence post, and used their observations to sketch the features of Mars. "It was an amazing sight," Roger recalls.

As he got a little older, he thought of being a musician, an inventor, or an engineer, or perhaps none of those but joining the Peace Corps. Instead, Mars came back into the picture. For his doctoral dissertation at the University of Minnesota, Roger became the first person to study actual samples of the Mars atmosphere in the lab, which he obtained by capturing minute quantities of gas from meteorites that had landed on Earth. His work contributed to the profound discovery that these meteorites had originally come from Mars.

Early in his professional career, Roger conceived the idea of continuing his Martian research by designing an instrument that would be included on a NASA spacecraft to Mars. One of his efforts was for a laser instrument that was accepted for a not-yet-named NASA mission. A year later, the committee that had made that selection for NASA was gone and the new one came up with another list that did not include the instrument Roger's team had proposed.

He came to recognize the difficulty of getting an instrument selected if it was brand-new and untested. For MSL, NASA headquarters had recommended that people or organizations interested in submitting proposals team up with others and submit jointly. Although an experienced investigator (he co-led the investigation by the Genesis mission to capture solar wind and return samples to Earth), being part of a veteran Mars team would greatly increase the chances of being selected. Roger managed to link up with another team that wanted to fly a panoramic camera, based on a previous version that had earned its wings by flying on Spirit and Opportunity.

The veteran team soon decided they were reducing their own chance of being selected if they partnered with newbies like the members of Roger's group. It looked as if Roger's team would have to go it alone, as Mars rookies.

A few years earlier, Roger had chatted with a French colleague, Sylvestre Maurice, about finding a way to work together on a joint international project. In Roger's eyes, Sylvestre was one of the up-and-coming French planetary scientists. Teaming up became more likely thanks to the French government, when Sylvestre and teammates were given a grant of about $1 million a year to develop a space laser in collaboration with Roger's team. The NASA Announcement of Opportunity for instruments to fly on a much larger spacecraft than ever previously sent to Mars was the trigger the two had been waiting for. Roger's team and Sylvestre's dived into writing a proposal for what they labeled the "Chemistry and Camera" instrument, intending that it be known by the catchy title "ChemCam."

The idea was this: ChemCam fires a series of laser pulses at a rock or a spot on the soil, from up to 25 feet (7.6 m) away. These have the energy of a million light bulbs focused to not much wider than a pencil lead. The pulses cause the target to emit tiny flashes of light, which are picked up by a telescope on the mast of the rover. From there, the light travels down through an optical fiber to the heart of ChemCam. There it is fed to a spectrometer inside the rover, which analyzes the wavelengths of the flashes, called "sparks." When the data is relayed to Earth, scientists would be able to determine the elemental makeup of the target material. That would allow them, little by little, to construct a geological history of Mars.

At that time, international cooperation was much smiled on in the US space community. In part that was because the United States wanted to share its leadership in space by getting other countries to take part, in the spirit of "opening our knowledge to others." In part, too, it was because countries like France were glad to untie their purse strings to help their own scientists and gain national prestige by establishing a foothold in the relatively new art of developing space hardware for a Mars rover.

"Frankly," Roger says, "I figured the chances of our instruments being selected were lousy. We had not yet proven we could build an instrument that would work on Mars. Our French partners hadn't, either." His hopes were dashed even further when he learned how many proposals had been submitted, and, worse, that only about six instruments would be selected.

In his book *Red Rover*, Roger tells about coming to work one cold morning in December 2004 and finding a voicemail message from the head of NASA headquarters' Mars program asking him to call back. Before he could, a press release in his email caught his attention. It announced the instruments that had been selected for MSL. ChemCam was one of them. Roger's one-sentence statement in his book captures his excitement: "We were on board!" The rest of that day was spent fielding phone calls from reporters and congratulatory emails, including one from the office of Roger's senator.

As you can imagine, creating a brand-new instrument to fly aboard a spacecraft that didn't exist yet and didn't even have detailed, finalized plans specifying size and design, is a formula for frustrations, misunderstandings, and turning gray at an early age. For Roger, the problems were not with the French partner team—that relationship went smoothly. Of the many challenges the ChemCam team faced, among the most frustrating involved a seemingly straightforward matter of an optical fiber to carry the light from the laser device to the ChemCam spectrometers. The JPL team insisted that the fiber have several connectors; otherwise it would be extremely difficult to remove and reinstall, which has to be done often during the process of assembling, testing, rearranging, and so forth that goes on throughout the later stages of getting a spacecraft ready.

Roger explained the problem this would create but the explanation didn't change anybody's mind, so he got on an airplane and flew to JPL. Meeting with the team, he patiently described why connectors are never used on optical fibers that carry light for a spectrometer. Unless the fibers inside the connector match up within thousandths of an inch, a great deal of the light being transmitted is lost. The JPL team said they had figured out a way to make the connections so that a full one-tenth of the light would reach the spectrometers. Or maybe as much as one-fifth. To Roger, that meant so little light that the ChemCam was likely to provide scant information of value to the scientists.

The ChemCam project was due for a routine NASA performance review. At that session, he found the reviewers were well aware of how eager the scientific community was for the data that ChemCam was expected to provide.

Roger received a phone call shortly after that meeting with welcome news: The JPL team had found a way around the problem. The optical fiber would have no connectors.

It seemed as if no sooner was one severe problem solved than another popped up, just as threatening to the whole project. But it wasn't all darkness and storms. By the fall of 2007, the instrument was coming together. The problems weren't all solved, but Roger was feeling confident about the progress.

Then a phone call came from NASA headquarters chief Mars scientist Michael Meyer. As Roger recalls the conversation, Meyer's first words were, "I have bad news for you. ChemCam is being dropped from MSL." NASA would not be able to provide the extra $2 million Roger needed to complete the development.

Mike Meyer at NASA headquarters and the team at JPL never wanted to ditch ChemCam. We all knew how important the instrument was to MSL and we had a hard time believing that NASA headquarters would really pull the plug. But Mike, acting under instructions, told Roger that the looming threat of cost increases for the rover, as well as for its instruments, was threatening MSL's overall survival, so cutbacks were necessary. Roger pointed out that his team had already spent $9 million of NASA's money and needed only an additional $2 million to finish the job. He told Meyer, "Some of the other instruments are more than $20 million over budget, and you're picking on us?" Mike pointed out that all of the NASA-funded instruments as well as the entire project were being asked to make do with the funds they had left, or face being cut.

Sometimes fate steps in. Just at this time, the director of the French space agency and his second in command were on their way to Washington to meet with the top people at NASA headquarters, and then traveling on to California to meet with JPL director Charles Elachi. The NASA people were encouraging, and Elachi promised to do his best to see that ChemCam funding was reinstated.

Meanwhile, although he was told he would not be receiving any more funds, Roger still had some NASA money left. Since he had never been expressly told to stop all activities, he kept his small team at work.

To help Roger in the face of his cost crunch, JPL had agreed to supply the electrical cables that would connect the French and American parts of Chem-Cam. When the cabling arrived, it was given a preliminary checkout, then attached to the instrument. As soon as ChemCam was turned on "something went very wrong." There was "a sickening smell of smoke," and his instrument went dead.

At JPL, in redesigning a pair of cables to combine them into one, an error had been made about which electrical circuit went to which pin. The result had been to short out ChemCam. For weeks, the team tried replacing one part, retesting, having no success, replacing another, retesting, with Roger despaired of ever getting the unit fixed. Two enormously frustrating months went by before it was finally working again.

Never think that the space program isn't political. Without any official notification, one day Roger saw a newspaper article saying that ChemCam had been restored on the instrument list for MSL. He was relieved, of course, but so were the Mars aficionados at headquarters. The plan to eliminate so much science within sight of the finish line had made no sense to them.

Roger's team had just gotten the engineering model working, and they found the data and pictures "amazing." In the following months, the Chem-Cam team received support and encouragement from many people in the space science community, which he found highly gratifying, and help from JPL engineers that was "enormously valuable."

A few years later, after Curiosity was at work on Mars, JPL Director Elachi was given the rare acknowledgement of being awarded the French Legion of Honor. In his remarks, he mentioned ChemCam, claiming that every time it zapped a rock, the buzzing sound it made was really a bit of French being spoken very fast: "Vive la France! Vive la France!"

The Mars Hand Lens Imager, the MAHLI (pronounced "MAH-lee") was one of three contributions to MSL from Malin Space Science Systems (MSSS). The founder and head of the company, Mike Malin, was also the principal investigator for the MARDI and MastCam cameras. MAHLI was the work of a team led by the company's Ken Edgett. Explaining his route to becoming a space scientist, Ken says that when he was a youngster, astronauts were walking on the Moon. That seemed to suggest an exciting, promising future in space exploration. In the fourth grade, to make his spelling homework more entertaining and to satisfy the "use each word in a sentence" requirement, he began writing short stories about the adventures of "Joe the Martian." That experience further stirred his interest in the planet, which made him want to know more and more about Mars. In 1975, when the two Viking spacecraft landed there, "the planet became a real place for me," he says. (Much later, keeping that boyhood character creation alive, Ken arranged to send Joe to Mars: A sketch of the character is aboard Curiosity, on the MAHLI Calibration Target.)

Ken calls himself a geologist, but he's not a geologist of the everyday variety. When pressed to be more specific, he describes his specialty as "geomorphology," which he explains as the study of landscapes and how they got that way. Ken's doctoral dissertation was on the sands of Mars and what we can learn about them from remote sensing.

I originally met him when he and a group of likeminded Mars geologists led a field trip to Washington State to show a group of us from JPL an earthly approximation of what the Martian surface is like, helping us understand the hazards that would be faced by Pathfinder's airbags and by its Sojourner rover. I found him to be happy, jovial, intense, and a master of his subject. I agreed with his view that beer and rocks go well together, voiced as we tackled massive floodwater boulders in the scablands of eastern Washington, climbing up a rock with one hand, the other gripping a Heineken.

Thermal infrared observations of sand dunes taken by Viking and the Mariner Orbiter had provided new knowledge about the size of particles that the winds of Mars can transport. The shapes of sand dunes turned out to be a clue to the directions of the strongest winds that blow through a particular area. Because the entire surface of Mars is subject to wind erosion and transport of sediment by wind, Ken says, "focusing on an aspect of how the wind interacts with the surface was a good move for laying the foundation for things we'd learn about Mars in the future, like how the landscape got to be the way it is."

In the fifteen years Ken has been at MSSS, he has worked on the cameras teams for a number of planetary orbiters. His first project had involved taking pictures with the Mars Orbiter Camera aboard the Mars Global Surveyor, and it was an eye-opener. In the process of selecting the targets, he noticed that what we now call Mount Sharp was built out of layers of sedimentary rock outcroppings. "I never knew before that there were such things." It was clear that these were actually rock. "This was a spectacular occurrence that changed our thinking about Mars," Ken says, because it meant the rocks had to be truly ancient.

Since a key part of Malin's business is building cameras for spacecraft, Ken and others in the company regularly talk about what they can do to improve the cameras currently being used. In 2004, the conversation focused on, "What kinds of cameras will consumers have in 2009, and what can we do to create something better than that?" Looking back, Ken finds this funny: "By 2009, the consumer camera was in your cell phone. We misjudged!"

Developing a proposal for a camera to go on the Mars Science Lab spacecraft was Ken's first opportunity as his company's number-one person on an instrument project. After months of creating a proposal that with appendixes can run to nearly a hundred pages, the months of waiting for a decision can be difficult.

How does a principal investigator learn that his or her instrument has been selected for a mission? Ken found out in an unexpected way. He explains, "I love this part of the story. They tell you that, if selected, you'll hear the news personally from a NASA headquarters official. My wife, Kim, was receiving NASA news releases by email and found a message with a subject line that said it was an announcement of the instruments that had been selected for MSL. Afraid to read it, she came and brought me to look at her computer screen. She said, 'Is this a good thing or a bad thing?'"

When he read the list of instruments that had been selected, which included his camera, Ken says he knew that the answer was both good and bad—good to be a winner, bad because a tough job now lay ahead. Both the head of Ken's company, Mike Malin, and Ken's graduate school advisor, Philip Christensen, had had instruments on many spacecraft, so Ken had seen first-hand the toll that the job takes, especially on the principal investigator, which is what he had just become. For him at that moment, it was a combination of exhilaration and "Oh, no—what have I gotten myself into?" He knew he would have little time to rest for the next several years.

One of the big technical challenges Ken found in the development effort was for the camera-focusing motors. They would need a lubricant that would work in extreme cold. Ken also found that the biggest challenge came from being new at trying to be a project leader. He says, "You're not sure which of the things coming at you are major problems and which are just minor. It's like when you're three years old and every little thing that goes wrong, every scrape or cut, every kid that snatches your toy away, is a major life crisis."

What does Ken Edgett think would be the most spectacular contribution his camera on Curiosity could make? "It would be signs of a fossil expression of a microbial community." In other words, capturing photographic evidence that life once existed on Mars.

"Not that I'm holding my breath," he says.

CHAPTER 10
Where the Devil Lives

B y 2006, MSL was still in what I think of as an awkward post-honeymoon phase. The high-level design had been completed, and the teams were now beginning the arduous, demanding process of working out the details—translating the high-level ideas into drawings, schematics, hardware, and software.

With the big-picture system design and operations concepts finally settled, you might think that the rest would be fairly straightforward engineering. But it was quite the opposite. This part of the work turned out to be where the devil lives.

The problem of cost continued to plague us in ways that no previous Mars project had to struggle with. In the 1970s, Viking, the first US Mars lander, was built at a time during the space race when money seemed to be no object. The cost was about $1 billion—more than $5 billion in today's dollars. For the next lander, Mars Pathfinder, twenty years later, we had been asked what we could do for something like a fixed budget of $280 million ($440 million in today's dollars). We did our homework and answered, "Not a lot, but we could put a lander and a little rover on Mars and operate it for thirty days, with a tad bit of science thrown in." We added that our degree of uncertainty in our cost estimates was greater than 30 percent. NASA headquarters said, "Great! Go do it." Our allocated budget included a 50 percent reserve. The mission was a big success, and we completed it under our budget, though with hardly a penny to spare.

Of course, that's what we're supposed to do. Yet the scope of MSL made Pathfinder seem like a little trifle by comparison.

That the Mars scientific community wanted to do MSL was clear, but the scientists didn't want it to be so expensive that its costs would push everything else people wanted to do off the table. From where I sat, it seemed as if the general opinion was that something like MSL should cost just a bit more than Spirit and Opportunity combined. Those twin missions together had a budget of $870 million in 2004 dollars—a bit more than that would be, say, around $1 billion.

Where did that "a bit more" come from? Was it derived by analyzing the scope of the project? No. It came out of what people *wished* it would cost—not recognizing that the designs for Spirit and Opportunity were largely adapted from the earlier Pathfinder mission, while MSL, with its much greater size, weight, and complexity, would have to be dreamed up almost from scratch. Yet that more-or-less $1 billion cost target was there the whole time, hanging over the small MSL team like a guillotine blade.

There is a lot of work needed on the design side before you can really come up with a reliable claim of a final cost estimate. On MSL, it took years to get even close. In early 2006, we were nearing the time of one of the most important project milestones: "project confirmation," when the science, the schedule, and the budget are all placed on the table for final acceptance by NASA headquarters management. After an elaborate bottom-up analysis, the JPL project management (by this time led by Richard Cook, one of the quickest thinkers I have ever met) estimated that putting MSL into space would cost $1.6 billion, which, as usual, included a healthy pool of reserve to deal with unexpected problems.

Outside of MSL there were howls of protest. At that cost, we would be draining funds being eyed by other science missions—projects that would have to be cancelled or delayed. Many felt that JPL no longer knew how to build something cheap; everyone remembered we had sent Pathfinder and our first rover to Mars for less than the cost of a major motion picture. Why couldn't we do that again? But we had known from the beginning that everything about MSL was going to be vastly harder than anything we had done before. It was obvious you don't build a rover five times heavier, ten times more complex, and requiring the invention of many never-done-before technologies, on a budget sized for something far simpler like Spirit and Opportunity.

In the NASA spacecraft business, frequent reviews of project status are a natural part of life. When you're doing projects that involve hefty price tags and engineering that has never been done before, it's natural that the people in charge of allocating the funds want to be convinced the people spending the money are making wise decisions.

Over the previous three years, the small MSL team had spent about 10 percent of its expected budget—mostly on contracts with outside companies developing needed technology. The teams had struggled in vain to find a design that would come in under NASA headquarters' rough cost target. Despite Mike Griffin's approval earlier in the year, officially MSL was still in preapproval status. To get the money needed to convert thousands of pages of PowerPoint files and spreadsheets into an actual detailed design of the whole spacecraft, MSL needed to successfully pass the next gate.

Across NASA, among the many reviews for flight projects are two major events that I refer to as the "monster reviews." They form the two most important funding gates. The first of those, the preliminary design review, or PDR, usually takes place about four years, sometimes more, before the scheduled launch. The team has to convince the review board that it has solid concepts for all major aspects of the mission. If you convince the panel that you're up to speed on all the major aspects, you then receive authorization and funding to purchase equipment, write contracts, hire contractors, and get down to detailed design and computer models from which real stuff can be built.

For MSL, this review was scheduled for June 2006, and I found myself in a somewhat awkward position. As the Mars Program chief engineer, I was asked to serve on the review panel, even though I had participated in MSL from the beginning as co-instigator of the sky crane concept, and I had peer-reviewed many of the early subsystem design details. Yet as a panel member, I was supposed to bring an independent, unbiased viewpoint, able to critique the design.

The review was held over four days in JPL's largest conference room, the Von Karman Auditorium, and only reinforced what I already believed. I had mixed feelings about how much work had been accomplished. On one hand, it looked as if there was a vast amount of work done on paper in the form of viewgraphs, schedules, budgets, and planning, but the project was then already about three and a half years from the 2009 launch date. The team really should have been further along in the details. Other than entry, descent, and landing, which had been the focus of so much attention for so long, I thought that the team was presenting designs that were rough, much less developed than they should have been at this stage.

At a PDR, the team is expected to present a final or nearly final design of every element of the spacecraft. A thumbs-up from the review panel says, in effect, "Your designs look solid and we consider that the project is ready to be funded for building the spacecraft." With a successful PDR, the project will then be "confirmed," and NASA headquarters will provide enough funding to cover costs until the project is well along.

One of the major elements for the rover was a complex robotic arm that would obtain a sample of rock or soil and move it to the instruments for evaluation. It was clear that the designs for the robotic arm were only at the early concept phase and were not anywhere near the stage where it could be fabricated and tested. The same was true for many other essential elements of the project. The avionics—the masses of electronics and cabling—were also far behind. The complex chips mounted on circuit boards were being custom-designed but the designs were still in the early stages.

Part of the reason for the lateness was that within recent months, reversing their earlier decisions, NASA headquarters and JPL upper management had decided that MSL was to have redundant electronics: If an item failed, a backup unit could take over. The plan up until then called for MSL to be what we call "single string": If any piece of avionics died, so would the whole mission. But with the price of MSL well over $1 billion, it was becoming harder and harder to justify flying without redundancy. MSL could be built to be just large enough to carry redundant components. So the decision to make most of the avionics internally redundant was reassuring, yet it threw a huge monkey wrench into the avionics design efforts, while also increasing the mass of the overall rover, which would make landing all that much more difficult.

At the PDR, many of the risks behind details like the impact of shifting from single-string to redundancy were only briefly discussed. But these were my friends and colleagues, people I knew were can-do types. MSL still had three years and five months before the launch date.

Making MSL seem all the more real, and creating an aura of confidence, was a full-scale wooden mockup of the rover that the team had built specifically to show off at the PDR. The review board members actually lined up for a photo op with it. If I had used my head, I could have come to firmer and less optimistic conclusions, but like the rest of the board, I was caught up in the excitement of MSL and the impressive mission-level details the team had presented. Our board essentially told them to "work harder" and catch up. If anyone could manage that, the board said, this MSL team could. I voted with the majority to give the MSL team a passing evaluation, knowing this meant NASA would start writing the big checks.

And that's what happened: In 2006, based on the report of the review board, NASA headquarters finally came through with the first round of big money for MSL. We were given a total budget allocation of $1.6 billion, everything that project manager Richard Cook had requested. This budget included what had been spent so far, plus everything we thought we would need to spend all the way through launch.

We would discover we were wrong by more than a little.

For all the engineers, technicians, and others hard at work on MSL, the thumbs-up from the review board hardly had an impact; they were too deeply immersed in their work to pay much attention. Consider just one example of the challenges the team faced, the problems of creating all-new tools. When a geologist out for fieldwork spots a rock or Earth layer that looks interesting, she reaches into her rucksack, grabs a rock hammer, chips off a piece, and brings it back to the lab. There she feeds it into one of her laboratory instruments, or several, and obtains the readings of the particular aspects she's studying.

The MSL rover was going to be a robotic version of that field scientist. The tools being developed for doing this we were referring to as "SA/SPaH," an ugly acronym—we pronounced it "saw spa"—that stood for the Sample Arm/ Sample Processing and Handling subsystem. The sample arm (SA), which in other contexts we usually called the robotic arm, would collect samples as either cores drilled from rock, or scooped soil, and then deliver the samples to a second, smaller arm (SPaH) that would process the material and ready it for delivery to the instruments.

The leader of the team, Joe Melko, had led his group through the design of a "corer"—a device that would remove a dime-sized core from a depth of about 4 inches (10 cm) inside a chosen rock. Each core would be placed on a tray attached to the SPaH, where a camera would capture images for relay to Earth. That core material—or, at other times, pebbles picked up by the arm's scoop—would then be crushed and sieved for delivery of a precisely controlled "aliquot," or sample, about the size of a finely ground baby aspirin. This would be dropped into one of the trap doors on the top of the rover, to fall into one of our two laboratory instruments.

When Joe had proudly showed a mockup and a computer rendering of the sampling process, I was impressed with the work. Not everyone was. One JPL review board member exclaimed, "This is the most complex system I've ever seen! There are far too many mechanisms involved!"

On the other hand, I liked that Joe had separated the coring and scooping functions on the arm from the sample preparation functions on the SPaH, despite the larger number of motors and gears involved. While each of these functions is complicated, I could envision them being designed and tested separately; this would allow the work to proceed at a faster pace than if we had to wait for a single, more complex mechanism to be designed and fabricated. I had to admit, though, that the criticism was valid in one vital aspect: If any of these many mechanisms failed on Mars, there would be little hope of getting the samples into the on-board lab instruments.

The verdict was that SA/SPaH be redesigned with fewer moving parts. Joe reluctantly agreed to form a "Tiger Team" to rethink the whole sample-handling system—simpler, and with fewer motors. (For those unfamiliar with the term, a Tiger Team is a group of experts pulled together temporarily to tackle one specific problem. The term is said to come from a 1964 paper in which it was defined, tongue-in-cheek, as "a team of undomesticated and uninhibited technical specialists, selected for their experience, energy, and imagination, and assigned to track down relentlessly every possible source of failure in a spacecraft subsystem.")

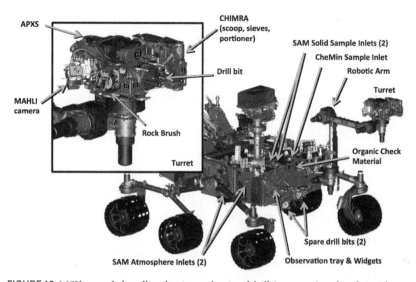

FIGURE 10. MSL's sample handling begins at the tip of drill (or scoop) and ends inside one of SAM's pyrolysis ovens or inside one of ChemMin's X-ray diffraction cells. The number and complexity of the processed surprised even the most experienced sample-processing engineers. (Courtesy of NASA and JPL/Caltech)

Though more than mildly irked, Joe good-humoredly shot a little video of his young son watching a child's Big Loader truck as it loaded and stored a pile of marbles in a remarkably complicated sequence of moves. Joe showed the video to all of us as a demonstration that the new assignment he had been given was valid: sample handling and delivery *could* be done with only a single motor.

In the new design, the smaller second arm was now gone and in its place was a complex arrangement of actuators and mechanisms. Well done, except that they would take up space that had been allocated for some of the scientific instruments. Worse, the robotic arm they would be mounted in had grown significantly in weight, now weighing about as much as my young daughter did at that time—about 60 pounds (27 kg).

As a small trade-off, the team wisely discarded the corer. The science advisors agreed it wasn't absolutely necessary to core the rock. The new arrangement instead called for drilling with a percussive device similar to the kind you can buy at Home Depot. MSL's version, though, would be a spinning percussive chisel to grind the rock into a fine powder. That left the challenge of figuring out how to capture the ground rock and transfer it into a device that could sieve it into fine grains. The solution came when they figured out how to design into the drill bits a fluted column; the grains would be drawn up the column and held there while the robotic arm completed the drilling. An electric vibrator plus the action of gravity would transfer the grains to what we started to call the Collection Handling for In-situ Martian Rock Analysis—CHIMRA, for short, which we pronounce "Kigh-MARE-uh." Mounted on the side of the drill, this device would sieve, filter and portion the powdered sample; then it would be moved to an access door atop the rover and dropped into the laboratory instruments.

(Despite the elimination of the SPaH, we didn't have the heart to rename SA/SPaH as "SA/CHIMRA." We like acronyms at JPL, but we like simple acronyms we can pronounce, so despite the design changes, the name SA/SPaH stuck.)

About this time, the dynamic and much admired SA/SPaH team lead, Eric Baumgartner, was drawn away to take up the position as dean of engineering at Ohio Northern University. By the time the design concept was finished, there was less than two years before the arm needed to be bolted onto the rover for the 2009 launch. There were still thousands of detailed drawings and schematics that had to be prepared, thousands of parts that needed to be fabricated, and many hundreds of hours of testing of multiple units before it could be ready for the testing of the complete robotic arm.

The job of doing the massive detailed design of this complex mechanical monster was handed to Louise Jandura. Gentle, sweet, and an effective team leader, Louise is one of a few experts in complex mechanisms at JPL. Still, this SA/SPaH was a challenge on a new, unprecedented level. For her, drill engineer Avi Okon, and their team, it would turn out to be one of the most exhausting jobs on MSL, and perhaps one of the most complex mechanical engineering challenges in JPL's history.

When the MSL team had committed to a landing choreography that included using a parachute as one of the slowdown techniques, we all anticipated we were taking on a tough challenge. We didn't even begin to understand how tortured an experience it would turn out to be. For the landing on Mars, the parachute, made of nylon fabric with Kevlar suspension lines, would need to open while the spacecraft was entering the Martian atmosphere traveling at supersonic speed, about Mach 2—something in the neighborhood of

1,000 mph (1,600 kph). Our spacecraft was going to be so heavy that we would need a *very* large chute, able to slow the craft quickly enough.

All of the parachutes used at Mars so far had been based on designs tested at supersonic speeds very high in Earth's atmosphere in the late 1960s and early 1970s, carried aloft by high-altitude balloons, with rockets to get the test vehicle and its parachute moving fast enough and high enough to reproduce the conditions at Mars. These were very expensive and complicated tests that have not been repeated since (although in 2014 we are trying again). Past Mars missions had been able to rely on the results of those old tests instead of running new ones.

Unfortunately, the largest of those tested parachutes was just a bit smaller than what MSL needed to land safely. Would MSL have to reproduce those expensive tests?

The answer was: "No, but . . ." We found we could continue to rely on the results of the old tests but we would still have to carefully test the strength of our parachute at sea level to ensure that once it inflates, it will not fly apart. The lead person on the parachute design and test effort, Doug Adams (his title was "parachute cognizant engineer"), booked time in the giant wind tunnel at a facility called the National Full-Scale Aerodynamics Complex (NFAC), located at NASA Ames Research Center in northern California. At 80 × 100 feet (24 × 30 m), the wind tunnel at NFAC is the world's largest, and it was the only one we could use, since we would be testing the world's largest supersonic parachute.

The strength test could not fully reproduce the conditions at Mars, when the parachute would be inflated high up where the air is very thin, even by Martian standards. Yet at the fantastic speed of about 1,000 mph (450 m/sec), the parachute feels terrific, hurricane-like forces when it opens. Our Earth test would inflate the parachute near sea level, where the air is one hundred times thicker and the speed of the air is one-tenth of the speed on the approach to Mars. Since force involved is a product of the density times the square of the speed, the effects more or less balance out: the forces on the test parachute when inflating would match closely enough the forces it would experience when inflating at Mars.

The first tests were not encouraging. The chutes inflated, but then ripped to shreds. The team decided the design needed to be strengthened, which

could be done by adding Kevlar ribbing—even though this was at the expense of making the chute heavier, in a situation where every pound counts.

When they tested the new design, a strange thing happened. While it was opening, the chute appeared to turn inside out and tore itself to shreds. That didn't seem to make sense. How could a parachute turn inside out?

Our parachute team had of course installed a battery of high-speed cameras filming each test. But the high-speed cameras were not high-speed enough: They had not captured enough frames of the action to reveal what happened.

For the next round of tests, the team loaded the test chamber with video cameras that operate at still higher speeds. The chutes shredded as before but this time the videos revealed what was happening. After weeks of hand wringing, analysis, and debate, we discovered that the parachute was filling up with air in exactly the opposite direction than it would on Mars. In the supersonic conditions at Mars, the parachute inflates from the bottom to the top like an umbrella. In the wind-tunnel test, we could clearly see that it was inflating from the top to the bottom, like filling a water balloon upside down. This backward inflation allowed enough time for the fabric at the mouth of the parachute to cross over and turn inside out.

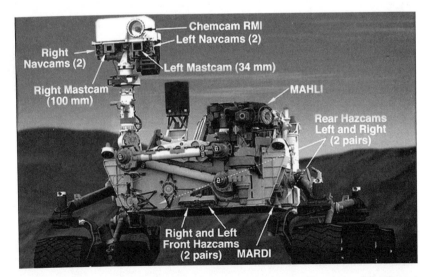

FIGURE 11. Locations of the seventeen cameras aboard MSL. (Courtesy of NASA and JPL/Caltech)

Happily, our worst fears turned out to be unwarranted. The condition would not happen on Mars even when inflated at twice the speed of sound. Another major concern put to rest.

By this time, the selection and placement of the various rover cameras was more or less set. In addition to the cameras for scientific purposes, the rover would carry twelve others dedicated to navigating the terrain and keeping Curiosity safe as she slowly moved from one location to another.

In the political structure in Washington, the funding for most JPL projects comes through NASA headquarters' Science Mission Directorate, so it was of high interest to us when word came in April 2007 that former astronaut Mary Cleave was leaving the post she had held for the previous three years as head of the Directorate, and a new leader, S. Alan Stern, was being appointed to replace her.

What would his priorities be? Would he continue the support for MSL and the other JPL projects, or would he campaign with NASA Administrator Mike Griffin to shift focus and spend the money on projects perhaps nearer to his own heart and scientific goals? Stern came with impressive credentials. Holding a PhD in astrophysics and planetary science, he had pursued research involving arcane topics with mysterious-sounding names like the Kuiper belt and the Oort cloud, as well as solar systems around other stars. He had been principal investigator on a number of projects with tongue-twisting names like the "ALICE UV Spectrometer for the ESA/ Rosetta comet orbiter."

We found out Stern's priorities in short order. He made it clear he was not happy that his predecessors had signed up for an MSL budget of over $1 billion. In many respects he was unhappy with the entire Mars enterprise. He felt Mars was getting too much of the NASA science budget while his own area of study, the outer planets, had not had a new mission in years. What's more, he didn't like the idea of projects that were not able to predict the future accurately enough to provide an up-front budget and schedule that they would be able to adhere to throughout the development. He would later be described in the *New York Times* as believing that "it is unfair to expect NASA to raid other programs . . . to pay for the [escalating] costs of current projects." And, hard words for us to hear, "The mission that makes the mess is responsible for cleaning it up." In other words, if you didn't accurately foresee all the challenges in developing the new technology that would be required for this project, don't come crying to me about it.

In June 2007, less than a year after the preliminary design review and less than two and a half years from launch, we gathered for the second of the two "monster reviews"—the critical design review, or CDR, held at the Embassy Suites near JPL and scheduled to last a daunting five days.

A separate review group was focused on cost assessments. Between the two groups, there were thirty-four reviewers—some from other NASA centers, some from the aerospace industry, some from university campuses, and others who had been nominated by people from outside NASA. In the eleven months since our first "monster review," the MSL team had grown from about 250 people to more than 600, and had been feverishly coming up with detailed designs, test plans, and test equipment, as well as yet-more-detailed cost estimates.

By this time, the other leaders of the project and I strongly suspected that the project would run through its $1.6 billion budget and might run out of money. The teams that needed to make decisions on how to stay within their dollar allocation had only so many variables they could control to keep the cost down. With MSL, the scope of what the team had to do was fixed, the schedule was fixed, and the cost cap was fixed. Our dwindling reserve was our safety valve.

That left only one other variable: Take more risk. Not more mission risk—that would be unacceptable. We could take some risk in our development plan, which would mean saving money if everything went right, but would place at risk our ability to be ready for launch in time if things went wrong. It would mean trying to get by with fewer test units, leaving us to perform critical design tests one at a time rather than in parallel. It would mean telling the hardware builders to skip building early prototypes for testing but instead go directly to the final design—which would mean trusting that every item would be close to perfect the first time. If every initial design were right, MSL would launch on time and under budget. It any initial design had a mistake that could not be corrected in time, then MSL would miss its launch window. We began spending more time crossing our fingers and hoping nothing in any of the great many designs would go very wrong.

The MSL team arrived for the CDR in the uncomfortable state of knowing that either the budget had to be increased, or they would be faced with knowing that any mistake could mean a two-year launch delay or, worse, possible cancellation. The head of the CDR panel, Orlando Figueroa, had directed the Mars Program at headquarters while Spirit and Opportunity were being developed. He asked that I serve as a member of their CDR panel, so as with the earlier PDR, I was to critique my colleagues.

The MSL team went in feeling confident. They had fleshed out the plans in exacting detail, with well-developed designs, and finely tuned simulations. But CDRs always take a toll. No matter how well prepared you are, people on the review board will always ask questions you never anticipated. The whole process is a harrowing experience.

The project management team argued that the project's budget was probably low by 10 percent. MSL was still budgeted for $1.6 billion, a third of which had already been spent. Perhaps worse, their documentation showed that the project was already beginning to run behind. Complicating the situation this time was the vast amount of complexity and detail that seemed to have come from out of nowhere in the past eleven months.

The team had proudly prepared plastic models of some of the key rover assemblies like the CHIMRA and the drill. I worried that the effort might backfire: Some of my review board colleagues examined the models and looked dismayed. They seemed to view the three-dimensional artifacts as making the designs look way too complex to be workable.

At the end of a CDR, the review group presents an evaluation of the project and offers one of three decision that are more or less the equivalent of, "Hey, these folks have developed a solid concept. They should go ahead—it's a sound investment of NASA's money." Or, "We find a lot of problems that haven't been satisfactorily solved yet. They need to go back and try harder." Or even the dreaded, "They need to abandon this effort and find something more useful to work on."

The project passed all of its review success criteria with two major exceptions. The board agreed that there were places where the design was not far enough along—in particular, with the SA/SPaH sample-handling mechanisms. The panel also debated at length on what to say about the schedule. Bobby Braun, the Georgia Tech professor who had served on my humans-landing-on-Mars panel, wanted to know if others were worried that the robotic arm and actuators might not get built on time. We all understood that the further behind the work became, the more expensive it would become to catch up.

Orlando asked if everyone would agree on saying that the project would probably end up needing an additional 10 percent.

I said, "Yes, but 10 percent may not be enough."

Bobby agreed. "They might need another 15 percent," he said.

The problem, everyone on the board knew, was that with Alan Stern raging against NASA's habit of allowing cost growth, it was the worst news we could deliver. It was unlikely that the project would get any increase; on the contrary, it was possible he would pull the plug.

After a day of deliberation, the board members had written enough that Orlando could prepare his report. He returned to JPL the following week and delivered the official CDR draft report to Richard Cook and the other MSL managers. It ended with the conclusion that "the budget reserves are inadequate."

JPL management was relieved that this complex, larger-than-ever design had received a thumbs-up to continue. But there was still Alan Stern to be heard from.

About two months after the CDR conclusions had been shared with NASA, word came from Stern: "I want you to descope your mission to fit the $1.6 billion budget we approved." "Descope": government-speak for cut it back, shrink it, make it less expensive.

Stern had read the report of the CDR board and decided that there was simply not enough budget to cover the cost of bad things going wrong, even if those bad things hadn't happened yet. But instead of finding NASA money elsewhere to augment MSL's budget, he told Richard Cook that the project needed to find ways to make up its shortfall on its own within the currently allocated budget, even if it meant stopping work on key science instruments or reducing or eliminating expensive tests.

If you were designing and building your own house and midway through construction you found that the cost estimate suddenly exceeded your budget by 10 percent, what would you do? You might go back to the bank and plead for more money, but if that doesn't work you have some hard choices ahead. Do you save money by telling the contractor to buy the cheapest roofing tiles instead of tiles that will last for decades? Not a good idea. More likely you'll look for cost-cutting measures such as dropping the deep stainless steel sink and going for good old reliable cast iron.

Unfortunately, from the very beginning, despite its "high price," MSL had been pinching pennies. What was there left to get rid of? We had already gotten rid of the stainless steel sink. We couldn't get rid of the parachute or the wheels. We already knew his view, made clear in statements and interviews, that he thought NASA had already done enough exploration of Mars

He knew that he didn't have grounds to cancel MSL outright. The project had not yet exceeded its budget, nor was there yet clear evidence this was inevitable—beyond his knowing Richard Cook's spare pool of reserve dollars was running low and that the report from the CDR board suggested the project needed budget elbow-room to deal with more surprises.

We were now just two years from launch. Project manager Richard Cook, under the descope command, once again had to examine every detail of the project, trying to decide what he could jettison to cut the budget by $60 million or so. After some sweating over the numbers and the places for possible cutting, Richard had to inform Stern there was little he could cut. There was one place, though, where Stern figured money could be saved: the rover's scientific payload.

The order to cut back or eliminate some scientific instruments carried a special meaning and special burden for one new member of the MSL team, John Grotzinger. He had earned a sterling reputation as a professor at MIT, and then at Caltech. At this time, he had just become the new project scientist for MSL, taking over from Caltech's Ed Stolper.

What is the role of a project scientist? John says it's someone who is supposed to supply "vision and leadership" to the scientists associated with the effort—those developing instruments and those working on the project in some other capacity. "But you don't get a rule book, and it wasn't obvious what I was supposed to be doing." He asked advice from fellow professors at Caltech and other universities who had served as project scientists on other missions, "but MSL was so different that their advice didn't help much." A major part of the problem, he says, was: "How do you do science with nine separate principal investigators?" So many more instruments than ever before carried to Mars, so many more scientists and instrument developers with questions, requests, and needs for guidance.

The first meeting John attended as project scientist was the session where he learned about the "descope" mandate. He was thrown into the fire, having to tell the instrument developers, "I now have to reduce your mission in some fashion." For Roger Wiens and his ChemCam, John was told about the phone call from NASA headquarters' Michael Meyer to Roger with the bad news that his project was out of money and was being cancelled. At least John hadn't been asked to make that phone call himself.

The more John learned about how each instrument was to be used, the more he realized that any descope of the instruments would be detrimental to the goals of the whole project. "I felt that any level of compromise would be unacceptable. I felt if I couldn't convince NASA headquarters not to cut back on the instrument funding, I was prepared to fight."

It would be a year before he worked out a solution.

CHAPTER 11
It's All About the People

This work is all about the energy, drive, and persistence of the people involved. If I thought what we had already overcome was like climbing Mount Kilimanjaro, I was about to discover that in fact we were a long way from reaching the peak of the mountain, and the rest of the climb was going to be far worse.

I was still Mars Program chief engineer, which meant that I was splitting my time between a lot of separate Mars efforts. I was still supporting Dara Sabahi's EDL team for the Phoenix spacecraft due to launch the following year, giving detailed design advice at MSL's subsystem reviews, and supporting long-term planning for the someday human Mars exploration.

Dara Sabahi shared with me his concerns about the MSL mechanical engineering work. He felt they were getting further and further behind. In his part-time role as JPL's chief mechanical engineer, he had been walking around and interviewing mechanical engineers on MSL to see if there were places that needed closer scrutiny. He suggested that I should do the same for other areas on MSL, spending time simply chatting with people to get a sense of how various team members thought things were going and if they were facing challenges that they were concerned about finding solutions for. Since Phoenix was looking in pretty good shape, I agreed, acknowledging that I thought I should spend a lot more of my time on MSL.

It wasn't long before Dara and I learned about one of the first big cracks in the machine. Adam Steltzner had called a meeting to inform us of some unsettling news about the viability of MSL's heat shield. All of our Mars landers require a heat shield to protect the lander or rover from the intense heat of entry into the Mars atmosphere. Back in the 1970s, researchers at Lockheed Martin (then called Martin Marietta) had invented a "thermal protection system" material, a blend of corkwood, epoxy, and millions of tiny gas-filled silica glass spheres, technically described as a Super Lightweight Ablator, with this specific version designated as SLA-561V. This material had been used on our Mars missions since Viking.

SLA has a unique quality. As the friction of the Mars carbon dioxide atmosphere begins to heat it, gases trapped inside the glass spheres slowly escape. This gas leaks into the oncoming hot flow of carbon dioxide, reacting with it and in the process stealing heat from around the heat shield before being blown away by the incoming airflow. The process creates a long contrail of hot gas in the vehicle's wake and leaves the heat shield much cooler than it would otherwise get. Without this process that we call "ablation," our heat shield and everything inside would melt.

When it came time to run the routine tests of our MSL heat shield material, the team put a "coupon"—a specimen of the material—into an Arcjet test facility. This is a hypersonic-thermal, ultra-high-temperature wind tunnel, able to produce an approximation of the surface temperature and pressure generated by a vehicle during entry into a planet's atmosphere. It's the only way we have of testing the material and design of the ablative heat shield. The full heat shield is never tested until landing day.

For our large lander, the heat shield would be the largest ever, some 40 percent larger than any previous one—nearly 15 feet (4.5 m) in diameter. For the first time on a Mars mission, our spacecraft would be using its heat shield like a wing to provide lift to help guide it to its landing zone, which would create far more aerodynamic forces on the material. That, plus the additional heating from MSL's extra-wide diameter, meant that its heat shield would be subjected to far greater heating and forces than ever before. In the Arcjet test, the coupon became so hot and the force so great that the hot air began to pull the heat shield material apart. The cork-wood/silica material began popping out of the honeycomb cells holding it in place. The heat shield was deteriorating. If this happened on the approach to Mars, it could not protect the rover, and the entire lander would become a fireball. Disastrous.

EDL team leader Adam Steltzner and some of the others suspected we were overtesting, that conditions on landing would not be nearly as severe as they had been told to test for. But no one could prove the worst-case condition wouldn't happen, so we had to use testing that would insure survival in the extreme situation.

It was back to the drawing board. With only two years to go before launch, they began a manic race to find a new material. Fortunately, researchers at the NASA Ames Research Center, while developing NASA's Orion program to replace the Shuttle, had been studying a new material called phenolic-impregnated ceramic ablator, or PICA, which turned out to be suitable for our need. But in order to use it, the team would have to invent a new tiling attachment and a new method for holding the material in place.

The news of the heat shield thermal material issue only confirmed what Dara had been seeing for himself. At his urging, I went in to have a conversation with my boss, JPL's Mars program manager, Fuk Li. He listened to my concerns and made a request that I only half expected. He asked whether I would be willing to take on the job of MSL chief engineer. He was offering me something I had wanted for a while. I felt my skills were a good match for what the MSL team was trying to do, and this was the next big thing in the Mars program. I happily signed up.

I knew, liked, respected, and admired the MSL team leaders and team members and looked forward to taking on the job. My only concern was that with yet another "leader," I would feel like a third wheel, or perhaps I wouldn't have enough to do. Those concerns didn't last long.

Fuk called a meeting with me, MSL project manager Richard Cook, his deputy John Klein, spacecraft manager Matt Wallace, and project system engineer Charles Whetsel. I was happy to hear that they were ready for me to come help. Although the honeymoon phase of the project had worn off, they were still quite enthusiastic about MSL, yet it was obvious they were starting to feel overwhelmed, especially over the struggle to find more things to cut as part of the descope effort that Alan Stern had demanded. They harbored hopes that I could dive in and find places where we could cut work or "reduce scope" to save money.

Shortly before the Christmas holidays, I moved into my new office in the overcrowded and rather dingy "temporary" trailers that were home for the MSL team.

On a spacecraft project, almost every day one team will discover that something another team is planning raises some kind of conflict issues with their one piece—the other guy's unit taking up too much space, or operating at too hot a temperature, or drawing too much power.

Complex systems are like an enormous painting. Hundreds of artists must each get their own little corner of the canvas. It is project management's and the systems engineer's job to make sure that each part of the canvas is allocated to the appropriate artist. When adjacent artists have widely different perspectives or palettes, the whole painting might have to be scrapped. It is the chief engineer's job to stand back, study the entire painting and resolve disasters in advance, which usually means working with the artists to solve the problem between them before the paint has dried.

For a spacecraft project, our canvas is far more multidimensional and the range of artistic skills extraordinarily diverse. We need experts on things you might not be surprised about such as artificial intelligence, robotics, motors, electronic circuit design, navigation, propulsion, software, and computers. But we also need experts in esoteric fields like tribology, cold lubrication, cabling and cable ties, material properties, radiation effects, explosives, fabrics, biosterilization, magnetic fields, battery chemistry, antenna patterns, flash-memory file systems, kinematics, fluids, hypersonics, nuclear physics, and many more. Somehow we need to find a way for this mind-boggling array of skills to come together so these big things we create will work they way they're supposed to.

I was to be MSL's art-studio repairman. But to be able to repair, first I needed to understand the whole of it. Despite having consulted on entry, descent, and landing all along, I knew little about the details of the inside of the rover, the instruments, and other many other elements. I had a vast amount of catching up to do. At first glance, it appeared that all the right things were happening. But the rover, descent, and cruise stages did not exist yet; the millions of components and parts and pieces (yes, *millions*) were scattered in the buildings of JPL, other NASA centers, universities, and contractors all over the continent and abroad. There were thousands of design details still not worked out, and all of these pieces need to be put together, first at the assembly, then the subsystem, and finally the system level.

For a mission due to launch in just under two years, this was absolutely frightening. We should have been at least assembling and testing the spacecraft subsystems by this time. The electronics test beds should have been fired up and working near flawlessly by now.

My first job wasn't to point out that the train was running behind schedule, but to find out what aspects were the furthest behind and what aspects were in most danger of never coming together at all. If there was work scheduled that didn't really need to be done, I had to find that, too, and get it off the schedule. The more I talked to people, the more I found work lagging behind where it should have been. I created a list of issues and started thinking through ideas for an action plan to tackle the problems that were slowing them down. I wandered around, checking in with old teammates, introducing myself to the people I didn't know, and picking up clues about emotional states and levels of confidence.

What I found, to a much greater degree than when I first made the rounds at Dara Sabahi's request, was a lot of frustration about the pressure to move very fast. Many wanted to get more people to help them speed up. Seeing where the project was at this point, pressure to go faster was understandable. But because of the budget restrictions, getting more people was understandably an uphill battle.

I had earlier learned a tactic I picked up from Dara. I had watched as he put into action a method he had sometimes used when his team was fraying at the edges. The situation, he had told me, isn't so much an issue of not enough time, but an issue of people. He said, "How do you get a group of people to come to an agreement when they have so many passionate but conflicting opinions?" His answer was, "*By getting them to listen to each other.*" His prescription went something like this:

"You gather all the key people in one room, telling them they may not bring their laptops, they may not bring their cell phones. You begin by going around the room, one person at a time, each speaking about what he or she says is the biggest problem for his/her team.

"While anyone is speaking, no one else is allowed to say a word. Not only that, no one is allowed to show a reaction of any kind—no smiles or frowns, no groans, no slamming of pencil or pen to the table. Each person *listens* to everyone else in turn, taking whatever notes they want. When you've gone all the way around the room, you start around the second time, and then you go around the third time. Why? Because the greatest frustration is, 'I'm not being listened to.' Here's a chance to put that to rest. For once, *everyone* is listening. There's a feeling of 'Finally I'm being listened to.'

"After the third time around, you go around again, but this time each person talks about what he thinks it would take to get things back on track. You can have each person write down everybody's suggestion, or better yet, you can make a list of the issues and suggestions on a whiteboard or a projected computer screen for everyone to see."

His final step is probably the toughest: "You get the group to arrange the items on the list in the order of priority, so you end with an action list that everyone agrees to. People leave the room with a sense of, 'They're taking me seriously.'"

"It works like a charm," Dara had told me.

He was right. I had watched him do this with the Spirit/Opportunity rover mechanical team when the airbags were being ripped to shreds and the team was deadlocked on how to proceed. He did it again when JPL and Lockheed engineers began losing confidence in each other in the early days of the Phoenix project.

Dara is a man you don't want to cross, but he has amazing group and team leadership skills. I had tried his method once before with a group from the Spirit/Opportunity team, and it had worked remarkably well. Now I was going to give it another try.

In part, this offered a way for me to find out quickly where the problems lay, before they became unrecoverable and threatened the project. I invited key members from several of the teams—but, intentionally, to avoid any sense of intimidation, no one from uppermost management—and told them to show up for an offsite retreat that Friday at a small conference room I had lined up at the old Ninth Circuit Court of Appeals courthouse in Pasadena.

When everyone had assembled, following Dara's technique, I started by going around the room. I asked, "What's bugging you? If you were in charge, what are the things you would do to make MSL a success?" And, "What technical processes do you think we need to follow that we're not doing now?"

A lot of the people began by saying how enthusiastic they were about being part of the project, how much respect and admiration they had for the people they were working with. What surprised me was that all of them ended up essentially telling the same story: Not enough time, not enough resources, not getting enough help, many technical problems that they didn't have time to hunt down and kill and that no one seemed to have answers for.

The more we went around the room, the more daunting it all sounded, yet the atmosphere seemed to get lighter for having the chance to sound off to the new Chief Engineer, as I sat there listening to them and taking copious notes. It was clear that they felt partly responsible even though I knew that underlying it all was a realization that this machine was far more complex and more difficult to design, build, and test than they anticipated. They simply did not have enough people or time to get it built fast enough. The complexity created a schedule problem that could not be solved within a budget that could not grow. What's more, I knew that the project management team and I were powerless to help them. We could not authorize spending money we did not have.

I asked if they had any suggestions for how the project could save money. A few people laughed, as if the question were some kind of bad joke. Others suggested that I ask project management to stop asking how far behind they were.

As the team filed out of the room at the end of the day, I could see that they had a spring in their step. They were being listened to. But I was feeling exhausted and daunted from what I heard. I began to feel that I had already let them down.

"Retreat" had turned out to be an apt title. They left the courthouse session feeling better. I left feeling as if a truck had run over me, as if I was carrying on my shoulders the burden of all their struggles. Feeling even more depressed than before, I wanted to run away. The only encouraging part was that I now had a spreadsheet of what the team thought they needed for them—and the project—to be successful.

I realized they did not have enough confidence in getting the work done in time to meet the launch date. I had to agree. Up until then I had been expecting that the team would have the self-confidence to overcome challenges just as we always had, and MSL would lift off the pad on schedule. The new input made me realize I had made a commitment we likely could not keep: the odds of launching in October 2009 were looking slim, even if the project were given the 10 percent additional funding that Richard had asked for. Without that, we had no chance of success. I began to feel that there was no way that all of these problems could be solved in time for a 2009 launch.

I went back to Richard Cook and the rest of the project management team to tell them about my findings. Clearly, they were caught in the middle. They were not happy, but there seemed little that they could do other than tell us all to press ahead and do the same themselves.

The threat of cancellation frequently haunted my thoughts.

Not long after, I arranged to meet for coffee with Gentry Lee, an engineer's engineer from JPL's upper echelon and a man I've always admired. He has multiple talents, he is eloquent, and he doesn't mince words; he's never been afraid of saying what he thinks. As chief engineer of Solar System Exploration, he is one of JPL's most senior engineers and had played a major role in the Viking program. But he had also earned renown as a writer, having coauthored several books with the late, celebrated science fiction writer Arthur C. Clarke. Gentry also wrote several sci-fi books on his own. To top it off, he shared credit with Cornell astronomer Carl Sagan for the 1980 television series *Cosmos*.

I had arranged to meet with him on the JPL mall next to the fountains, a place we could have coffee together and talk without being overheard. I knew that as part of JPL management, he likely shared management's view that MSL was running a bit behind schedule but was not in any serious trouble. After the greetings and a brief recounting of my off-site meeting, I said, "Gentry, I give MSL's 2009 launch a 5 percent chance." He weighed that, sipped his coffee, and said, "Rob, that's good. What's bothering you?"

"No, Gentry," I said. "Only a 5 percent chance of being able to launch on schedule!"

Gentry spit out his coffee all over the table.

I wasn't sure which of us was more embarrassed. He looked shocked by my statement, his face turning pale.

"Are you *sure*?" he asked.

"I can't prove it, but, yes, I'm quite sure," I answered. "There's just way too much to do to be ready to launch in 2009, and if we did, I'd be terrified. I wouldn't trust the vehicle."

He sighed, took off the cap he always wears, and rubbed his forehead. After a few moments, he said, "You just can't tell management what you just told me. You have to have facts, you have to have metrics to show them." He went on, "Without facts, you're not going to make a case that 2009 isn't possible."

I squirmed. "The downside of going for 2009 at this breakneck pace is that people will make mistakes, and there will not be enough time to catch and fix them before it's too late."

After another thoughtful, painful moment, he asked, "Can you put the facts together?"

"I'll try," I told him. "But I can't think of any piece of data that would be convincing."

"Keep trying," he said. "Go ahead and let your concerns be known, but if you want decisions to be made, you will need defensible data." That was as much encouragement as he had for me, but I knew he was right.

I went back to speak in more detail with project manager Richard Cook, telling him about my conversation with Gentry. He suggested that I talk to Fuk and to flight projects manager Tom Gavin. They gave me virtually the same message, the same one that Gentry had predicted: "If you can come up with metrics, bring them and we'll talk." Plus various versions of, "Until then, your job is to solve the problems as best and as fast as you can." Tom added a word of caution. "Be careful about conveying downer messages to the team. It's likely to make them want to give up."

Repeatedly I tried to come up with metrics to show that the best course would be to stop, regroup, and go for 2011, but I couldn't land on a single item that would be accepted as proof. I began to realize my dilemma. The burden on a project has traditionally been to try to show that you *can* make it even if some things go wrong. I was being asked to prove the opposite.

I had to accept reality. As long as the schedule showed that launch in 2009 was possible (even if unlikely), the only thing we could do was simply plow ahead at full speed and hope to hell we did not make any unrecoverable mistakes.

Despite my anxiety, I was left to go back and tell the teams I had no answers. They just needed to continue to work as hard, as smart, and as fast as possible. It may have been the right thing to do, but still I felt I had let them down.

CHAPTER 12
Bad to Worse

As bad as things were looking, I worried that without additional money, the situation was about to get much worse. Instead we were greeted with totally unexpected news. A NASA-wide email announced that Alan Stern, after less than a year at NASA, had resigned.

I had always had high respect for his credentials and his commitment to keeping costs in check, but Stern had never been a fan of Mars exploration or of the Mars Science Laboratory. One anonymous insider (from NASA headquarters, I presume) posted on the web, "Staff reaction to the resignation was I think a bit of a shock and a great sense of loss. Alan brought a breath of fresh air and a new spirit to the organization that was a lot of fun." It may have been fun there, but given our continual efforts to pinch pennies and find ways to save money at the expense of getting the job done, "fun" was the last word I would have chosen to describe our experiences.

Still, I wondered: Was his departure going to bring about changes in our too-few-people, too-many-problems, not-enough-time-or-money dilemmas?

The answer came soon enough. Word arrived from NASA headquarters that Mike Griffin had hired Ed Weiler to replace Stern as the associate administrator for science. Ed had been in that job before and knew how hard it was to pull off something like MSL. Within a few months, NASA headquarters provided Richard Cook the additional 10 percent he had asked for a year earlier. If we thought this was a solution to our schedule problems, we would soon find ourselves gravely disappointed. Still, for the moment, it was welcome news that gave everyone from the newest hire to the JPL director a renewed sense of confidence in the project.

We were able to add another hundred people onto the rolls of the MSL workforce, in addition to the seven hundred already on the books. Still, for a time adding extra people to help us go faster became a bit like hiring nine women to make a baby in one month. Instead of speeding things up, it actually slowed us down as we struggled to train the new hires into the workings of this complex system. It's not like a contractor hiring a new carpenter who, if he or she is a good, experienced worker, can step right in and start sawing and hammering. In a business like ours, it's not anywhere near that easy. For a time, productivity dropped.

Though the project was still struggling, these steps took an enormous amount of pressure off all of us. At last we were allowed to focus 100 percent on trying to get to the launch pad by 2009, though even with the new funding, it still seemed far less than a sure thing.

By May 2008, we were able to begin the assembly process for building our spacecraft, a process we call ATLO, for assembly, test, and launch operations. Most of the ATLO would take place in JPL's High Bay 1 clean room, the same room where many of our earlier spacecraft had taken shape. For MSL, the ATLO clean room had to be a super-clean environment, with the air constantly filtered and kept clear of particles. It is the place to be on a hot smoggy day or on a day filled with summer pollen. We keep it clean in order to keep Earth bugs and other contaminants away from our rover so that we won't populate Mars with bacteria from Earth. Before entering the clean room, we get decked out in what we call a "bunny suit"—cap, face mask, gown to mid-calf, trouser coverings, gloves, shoe coverings, all in white, with nothing showing except the eyes. (It's surprising how you come to recognize people just by the body shape and the eyes.)

About three blocks away are our "test beds," laid out in two separate rooms. One has a floor covered with rocks, sand, and gravel to simulate the surface of Mars, and large bright lights that simulate the sunlight of Mars; this will be home for our Vehicle Systems Test Bed—a fully functional engineering model of the rover, where all of the rover's surface functions would be tested and where the rover would be put through its paces. This test bed rover will never fly to Mars, but we can get it dirty, we can put it to the test of drilling rocks, and we verify every aspect of the millions of lines of software.

The adjacent room, visible through a glass wall dividing the two areas, is a quasi–clean room area with a raised floor that we call the Mission Systems Test Bed. Laid out are all of the electronics to make a complete working space vehicle including the brains of the rover, but it's a space vehicle without a body: The components of the rover, the descent stage, and the cruise stage are spread around on tabletops and workbenches, with masses and masses of thick black cables that connect the separate elements and also connect to tall computer racks all around. It's a cluttered atmosphere, and rather uncomfortable because constantly running fans make a nonstop high-pitched noise. This area is the focus of testing all phases of the MSL mission, to ensure that the pieces all work together. No coffee or food is allowed in these rooms, and no one works here alone; a buddy is always required in case of injury, fire, or other emergency.

Ordinarily, as each piece of the subsystem hardware arrives at JPL from its fabrication facility, it's put through rigorous testing in small labs before being sent to the test beds. No one expects these first "prototype" or "engineering" units to pass perfectly the first time. Once the appropriate test team or teams have noted a problem, the item is fixed on the spot or sent back to the contractor or developer for rework. The "flight unit" is then built and put through another round of rigorous testing in the same small labs. Only after the unit has passed the testing and been declared ready in every way for flight is it sent to the ATLO High Bay to be cleaned and become part of the buildup of the flight rover.

Unfortunately, the subsystem teams were running late. To keep the ATLO team on schedule for launch, instead of waiting for the final flight units to arrive from the subsystem labs, the project asked that subsystem teams and contractors deliver whatever hardware they had on hand, even though in most cases we knew it wouldn't be the flight units. They did; in some cases this left them without hardware to test to prove out the final designs. They—and we—would have to risk that the flight units they were building would be free of bugs and design errors. If there were bugs in the design, the ATLO team might be the first to discover them.

(A revealing and amusing time-lapse covering months of ATLO activities can be viewed at CuriosityRover.info/ATLOvideo. Note that not until near the end of the video—a few months before launch—is the rover built up enough to begin looking like the Curiosity you see in pictures from Mars.)

Once the buildup started in the ATLO clean room in the early summer, I needed to add an ATLO visit to my early-morning daily rounds. About 7:00 or 7:30, there would be a "tag-up"—a twenty- or thirty-minute meeting with the night shift just getting ready to leave and the day shift just arriving, as we all listened to the recitation of what went right overnight and what went wrong.

At first, the process seemed to be going well, though slowly. But the ATLO team began to find many little mistakes that would normally have been caught before the hardware was delivered. Usually they were simple mistakes—for example, finding a signal that should have been on pin 23 of a cable connector was instead on pin 32. As these problems were found, the ATLO team would log them in a problem/failure report, with a number assigned, and listing the name of the person who was going to address the issue. The team building the rover couldn't stop to fix things. They had their hands full just trying to keep up with the job of assembly. Repair and retesting would have to wait.

Adam Steltzner and his entry, descent, and landing team had continued doing a stellar job. I was able to enjoy the time I spent working and watching his group continue arriving at innovative solutions for this novel landing approach we had conceived.

On the other hand, we were living with what I considered to be a dirty secret. The EDL team had not been testing the inflation of parachutes in a Mars-like environment. While we could test the strength of the parachute within the setting of a giant wind tunnel, this was in low-speed conditions, at sea level in a thick Earth atmosphere.

Worse, we hadn't seen the actual moment when the parachute began to inflate other than in films from the halcyon days of the 1970s, when NASA had spent hundreds of millions of dollars testing parachutes miles above the White Sands Proving Grounds in New Mexico.

Although often batted around as a "Wouldn't it be great if" scenario, we couldn't afford to set up tests in the conditions that would allow roughly duplicating the speed and air density that our parachute would experience on the descent to Mars, nor did we have movies or images of parachute inflation taken at Mars from our past probes. That scant data from the early 1970s was all we had, and it told us we could expect that our new extra-large parachute would probably be able to open as intended. *Probably.*

The Phoenix spacecraft developed by Lockheed Martin, JPL, and the University of Arizona was due to arrive at Mars on May 25, 2008—barely a month away. Phoenix would be using a parachute that was smaller but identical in design to the one we were counting on using with MSL. If the landing went smoothly, it would be a great reassurance that the EDL team from Lockheed and JPL still had the "right stuff" to set a lander safely on Mars.

But the more I thought about it, the more I worried about the opposite: What if something went amiss in the Phoenix landing? What if it failed and it appeared the parachute might be at fault? It would put the brakes on our MSL super-chute for sure. If something went wrong, how would we know whether it was something else that was at fault, and not the parachute?

The Lockheed team came up with the answer. NASA has that pair of Lockheed-built orbiter spacecraft traveling through the skies above Mars—Mars Reconnaissance Orbiter and Mars Odyssey. Their route around the planet, and what their cameras are pointed at, are under command from Earth. (In fact, some of the "Ops team" members sit at consoles at JPL; others are at Lockheed in Denver.) The Lockheed team thought it was crazy not to try to take a picture of Phoenix lander as it descended. Could one of the orbiters really take a picture of the parachute?

I went for an urgent talk with the two Lees—Fuk Li and Gentry Lee—and made the case for setting up the Reconnaissance Orbiter so it could take a picture of Phoenix while the chute was open. But it wasn't their call, and the Phoenix project management thought it was too risky. We would be trying to take pictures just when the orbiter was also busy recording critical data from Phoenix to be played back later to JPL. A mistake made while attempting the picture might ruin the orbiter's ability to listen to what Phoenix had to say.

JPL management wanted to see data to establish that the pictures could possibly show something of value. At hundreds of kilometers away, could the camera make out details of the parachute? Yes: We simulated the image and found that the parachute would be clearly visible.

But would the parachute and lander even be in the picture's field of view? That was harder: there was a good possibility that the orbiter's camera, which

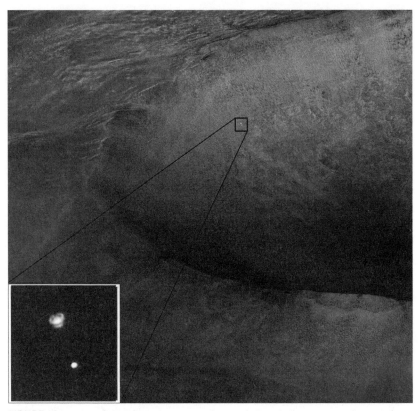

FIGURE 12. Image of the Phoenix spacecraft as it descended on its parachute in front of Heimdall Crater in May 2007. (Courtesy of NASA, JPL/Caltech, and the University of Arizona)

of course would have to be programmed in advance, might instead take a picture to early or too late, or that wasn't pointed accurately enough to include Phoenix. After sweating over the math, the MRO team and I were able to prove there was a better than 50/50 chance of getting a single good picture timed to capture an image of the parachute, an image that might just turn out to be crucial to the success of MSL.

In the end, the decision had to go all the way up to NASA headquarters' Mars Program director. Though some at JPL still felt it wasn't worth the risk, we got a "go" just two weeks before the arrival of Phoenix at Mars.

* * *

Phoenix arrived, and its landing was flawless. The picture I had arranged for was, after all, not needed. Nonetheless, the photograph captured by the Orbiter turned out to be dazzling. Never before had one spacecraft observed another descending to the surface of a planet. The picture has become one of the iconic photos of the space age.

Just days after landing, Phoenix sent back images showing that a few inches below the dusty surface of the northern plains was a vast reservoir of frozen water: ice. Finally we had actually seen evidence of "water" on Mars. Not only water ice, but salty, briny ice surrounded by what appeared to be perchlorate in the soil, a chemical commonly found in solid rocket fuel that, though toxic to humans, serves as food for some microbes. What did this mean? We still don't know, but the news excited scientists and the space-absorbed public around the world.

Phoenix continued to carry out science experiments for five months, finally growing silent as expected when the oncoming Martian winter no longer provided enough sunlight each day to recharge the spacecraft battery. This is what people in the space business long for: a spacecraft that carries out its mission successfully for its full planned lifetime.

At that moment in the MSL project, I would gladly have settled for just getting our beast to Mars and having it land safely. But the problems on MSL weren't getting any easier. I work surrounded by people who share various strengths, one of which to me is predominant: ingenuity. One place that required a lot of ingenuity was figuring out how to operate a rover on about as much energy as it takes to light a 100-watt bulb. On a typical busy afternoon, Curiosity would need 500–1500 watts to operate, but Curiosity's power source was capable of generating only a little over 100.

How do you run a 1500-watt rover on only 110 watts? All of our rovers have required more energy to operate than they could generate at any one moment. On Pathfinder, Spirit, and Opportunity, we had developed a lot of experience designing rovers that had to operate with very low power. The problem is solved by making our rovers what I refer to as "narcoleptic": They are awake only for a few hours out of each sol, each 24-hour-and-39-minute Martian day. When the rovers are awake, they would use more power than their solar array's power supply could deliver. The rover would have to operate from its battery.

We came up with a design that put the rovers to sleep at the end of the day's activities. When the rover is asleep, its computer and most of its electronics are actually turned off. A typical Mars day has the rover waking up around 9:00 a.m. Mars local time, just before for the daily set of instructions that cover the next sol or the next few sols arrive from Earth. It would usually be given a few short activities before taking a nap for a few hours, and then would wake again around noon after the sun or a heater has warmed up its mechanisms to a nice toasty 30 degrees below zero.

The next two or three hours would be when the rover does its heavy work of the day. Then it would close down for another nap until 3:00 or 5:00 p.m., when it would wake for twenty minutes to send its data to one of the Martian orbiters for relay to Earth. Then it would go back to sleep. A day in the life of a rover is a bit more like an old dog than a racecar.

We had known from the outset that MSL would not be solar-powered like all of our previous landers. At the landing sites that my scientist friends were ogling, solar panels would not have provided enough power to last. No, we would have to bring with us from Earth enough energy to last at least one Mars year.

The choice for a power source wasn't difficult. We would use an evolutionary version of a design produced for NASA by the Department of Energy that had been relied on by planetary exploration missions for more than forty years. Called a radioisotope thermoelectric generator, or RTG, it would have as the heart of its power source 10.6 pounds (4.8 kilograms) of radioactive plutonium dioxide. This plutonium gets hot, producing 2,000 watts of heat, of which about 95 to 110 watts go into producing electricity. Unlike solar panels, an RTG would be capable of providing that power all day, every day, for years.

MSL's 110-watt electrical power supply was far smaller than what this massive rover needed, even with frequent snoozes. To get it warm enough to operate even by noon, we would need a lot of heater power. The MSL thermal team found a way to steal enough energy for keeping the insides of our rover from becoming too cold throughout the Mars nights. The key was a scheme for capturing some of that 2,000 watts of waste heat generated by the plutonium power system. They did this by putting "catcher's mitts" on either side of the hot container of plutonium and running tubing that passed through the mitts. Liquid Freon pumped through the tubing would pick up heat from the power system and carry it inside the body of the rover holding the electronics and sensitive instruments.

When all the components were assembled, we found that this solution worked brilliantly. We could now be confident that the rover body would stay near room temperature for as long as the plutonium and the pumps lasted, putting our MSL rover in great shape to survive through the long Mars winter. (A Mars year is about twice as long as an Earth year, so the winters are twice as long as well.) Still, there was a downside: This Freon would only keep the rover's body warm. It would do nothing to keep the rover's extremities from freezing. For that we would have to use the rover's precious electrical power.

I wish every design for MSL had worked as well as that one. Though there were hundreds of other success stories like this, there were still enough frustrations to stop us cold. Just ahead we would run into one major problem after another, any one of which seemed severe enough to put us out of business.

The key to discovering design problems is testing. We have great design engineers, but what looks like a good design, even what looks like a *great* design, is only as good as our imaginations. Testing can reveal huge holes in our thinking. The number of different conditions that need to be tested is almost inconceivable: it's in the *millions*. Given our limited test resources, shaking out the designs required every moment of testing time. Especially because we were running so far behind, the competition among the teams to schedule time in the test bed became intense. Test bed time slots were assigned around the clock, seven days a week in three eight-hour shifts per day, which might be 8:00 a.m. to 4:00 p.m. on a workday or midnight to 8:00 a.m. on a Sunday.

That wouldn't seem quite as bad if the team could walk in and quickly begin testing. No, it could take four hours just to load the software required for the testing, and might take another hour at the end of the shift to shut down and be ready to leave when the next team showed up. It wasn't uncommon to put in eight hours and get only two or three hours of actual testing done. For those engineers using the test bed, the experience was often frustrating and always tiring. Day after day of long shifts with only occasional food or drink in a glaring white lab, surrounded by whining equipment, can sap your soul. Yet every one of these people had struggled and fought against the odds to be hired by JPL, and they felt themselves blessed when they were accepted into this exclusive "club."

The scientific community wanted to take it for granted that there would be sufficient power for their instruments, and that the rover would provide temperatures the instruments would be comfortable in. Getting the laboratory equipment to Mars was only the starting point of the adventure, and would mean worse than nothing unless the sample-collecting devices and instruments worked once there. As for the temperature challenge for equipment outside the rover body, the final specifications that our thermal team had been given called for it to be able to perform all actions in conditions as cold as −184 F (−120 C).

On Earth, we have little need for motors and gears that can operate in extremes of cold. If you need equipment that can function in conditions where the temperatures get down to, say, −58 F (−50 C), you won't have too much trouble finding them. But few off-the-shelf parts will work in the extreme cold of Mars unless they are electrically heated first.

This called for more of the hands-on problem solving at which JPL teams are so good. The mechanical team designed a new type of motor and gearbox called a "cold actuator" that could operate at very low temperatures without the need for heating. To save weight, the mechanical team chose to have the needed motors and gears fabricated out of titanium. At these temperatures, any liquid lubricant would get thick and heavy, rendering the actuators unusable. Instead, the team found a type of dry, powdered lubricant called molybdenum disulfide.

Well before MSL became a real project, the actuator team had fabricated a number of these cold actuators, and they tested well. It looked like the various actuators needed in the robotic arm, as well as in the mast and wheels, would operate just fine without being heated. With only 110 watts of power from our power supply, this solution was absolutely mandatory. Problem solved.

Earlier, I had agreed with the team's decision to put off doing a life test on these cold actuators until we had accumulated enough of them to make the testing worthwhile. When that time arrived, the units were put into a cold chamber set to duplicate the cold, dry, nearly airless conditions on Mars, where they were left to spin and spin and spin. How much testing would be required? To last at least a Mars year, we would need them to continue turning for many millions of revolutions.

The tests started off well, and the reports were accompanied by smiles. But the smiles didn't last. After a few million revolutions, the teeth on the gears began wearing, cracking, and falling apart. At these temperatures, the combination of dry lubricant and titanium gears wasn't working. This brilliant-sounding cold actuator approach was unusable, after all.

Back to the drawing board. We had to revert to the tried-and-true wet lube and "warm actuator" method we used on the earlier rovers. The team would have to go back from titanium to stainless steel, but this created another significant problem. To use stainless steal at cold temperatures, we would need to provide more space and more "slop" between the gears so that the wet lube could fit in the spaces between the teeth. This meant that, in some positions of the arm, we would not be able to rely on it to stay put once positioned above a rock.

Worse, these new actuators would not be able to operate at frigid temperatures. We would have to add separate heaters for most of the rover's thirty-two motors and gearboxes—five for the arm, four for the drill, ten for the wheels, four for the CHIMRA instrument, one for a dust-removal tool, three for the covers protecting the inlets for sample material being fed to the instruments, three for the mast, and two for the high-gain antenna. This doesn't include the motors in the cameras and other science instruments.

The thermal team would have to provide each actuator with its own electric heater strips wrapped around it. Yet because these actuators weren't designed to be heated, they didn't offer flat surfaces that heating strips could be bonded to. There was now not enough time in the schedule to redesign the actuator housings to be efficiently heated. To get the motors and gearboxes warm enough in the late morning after a cold night on Mars, we were forced to design and install grossly oversized electric heaters that were far less efficient than what we could have had if we had gone this route from the outset.

Each of those heaters would of course need electricity. Lots of it. Throughout the entire design process, electric power was one of the big, recurring headaches. Even with our schemes to put the rover to sleep often and steal heat from the RTG power supply, there still wasn't enough power. By the time the rover warmed up those big, heavy actuators with their big power-hungry heaters, it would be noon and the rover would be ready and raring to drive or drill—except that the battery would be depleted. It would have been nice to use some of the extra heat from the power generators, the RTGs, but we could not find a way to provide all the plumbing for carrying the heat to all the moving parts.

If the battery were twice as big, it would take twice as long to charge, but at least Curiosity could be sent out for a drive, or to drill a rock, or to scoop some soil and drop it into instruments for analysis.

It never ceases to amaze me how often the solution to one problem creates a new one somewhere else. A bit like a paper bag that has been filled with air, you punch one side of it and a hole pops up somewhere else. Those new heaters had added a huge, unanticipated burden on the whole rover. What now?

By coincidence, about the time of the bad news about the power problem, we got an email from our friends at NASA's Goddard Space Flight Center who were building the Sample Analysis at Mars (SAM) instrument. They had been testing to see how much power would be needed to warm up their little internal ovens that would cook the Mars rock and soil samples. The message mentioned that their instrument might need about 1200 watt-hours to do a day's worth of science.

That was about *four times* more than they had been allocated. They had been harboring hopes that we could somehow deal with it. Before they pulled their instrument apart and struggled with finding a solution, they wanted to see what our reaction would be.

I always try to remind myself that it's important in these situations not to panic. But between SAM, the actuator heaters, and other surprises, we were suddenly faced with a situation that appeared to have no solution. Our battery was simply not big enough. There was not enough room for adding a second battery, as well as no time to have a new one designed and fabricated.

With a little over a year to launch, we were out of power. The problem threatened to sink the entire project. Our only hope now was to try to make a convincing case that we could do most of our mission in the heat of the summer when it was warm enough to avoid using the actuator heaters for long.

In my thirty-plus years designing and building spacecraft, I am constantly amazed at how often a big project can be undone by something as small as gears or ovens.

We happily received one big boost to morale, and it turned on a question people often ask: How do our spacecraft get their names?

For the Pathfinder mission in 1997, the manager of the Sojourner team, Donna Shirley, came up with the idea that the honor of naming the rover should go to a student winner of a national contest, with the entries being judged not just on the name suggested but also on an essay the entrant would write to explain why the name they were proposing was appropriate.

Entries were collected over the period of a year and submissions came from around the world. The winner of that first contest was Valerie Ambroise, a twelve-year-old from Bridgeport, Connecticut, who suggested naming the rover in honor of Sojourner Truth, an African-American who during the Civil War had campaigned intensively that all people should be free and that women should have equal rights with men. We were very proud of that name.

The idea of a contest for schoolchildren to propose names for our rovers became a tradition that NASA has followed ever since. After the MSL rover contest was announced, NASA headquarters received some nine thousand entries. Following a prescreening, a selection was sent to JPL for a final recommendation by the MSL team members. To me, the choice of "Curiosity" was a no-brainer. The winner was another twelve-year-old sixth-grader, Clara Ma, from deep in mid-America: a town in northeastern Kansas with fewer than fifty thousand inhabitants.

Her essay read:

Curiosity is an everlasting flame that burns in everyone's mind. It makes me get out of bed in the morning and wonder what surprises life will throw at me that day. Curiosity is such a powerful force. Without it, we wouldn't be who we are today. When I was younger, I wondered, "Why is the sky blue?" "Why do the stars twinkle?" "Why am I me?" and I still do. I had so many questions, and America is the place where I want to find my answers. Curiosity is the passion that drives us through our everyday lives. We have become explorers and scientists with our need to ask questions and to wonder. Sure, there are many risks and dangers, but despite that, we still continue to wonder and dream and create and hope. We have discovered so much about the world, but still so little. We will never know everything there is to know, but with our burning curiosity, we have learned so much.

I found that essay stirring. It gave me renewed pride in being part of the project, and our rover had a name that could not have been more fitting. "Curiosity" was squarely at the heart of what MSL was all about.

Continuing a practice that started with Sojourner, we would often speak of our Curiosity rover in the feminine: "she."

CHAPTER 13
Up Against a Brick Wall

As summer season approached, few us of found time to take off. For those who did take a few days, it was a most welcome break. Working on flight projects at JPL is not a nine-to-five job, and never has been. That's true for almost everyone. It's especially true for a chief engineer.

In those days I was getting up around 5:30 every morning and arriving at the lab about 7:30. I would find my electronic calendar totally filled, often with ten or even twenty scheduled meetings in a single day, most of them conflicting or overlapping with others. At the first opportunity, I would check my computer to see the report of what tests the crews that had been working through the night had accomplished. Then I'd log into the problem reporting system to see what issues had come up. The teams would have handled the small issues on their own. Serious issues that required coordination or some sort of reworking would be reported in one of those problem/failure reports (PFRs). On MSL, there were typically four or five new PFRs each morning and another handful throughout the day; at our peak, I would see that number rise to about twenty to thirty per day. I kept my eye out for problems that crossed subsystem boundaries or those that went beyond the capability of the particular subteam to fix on their own, meaning they would need help in finding a solution.

Next I'd look at the morning batch of emails, always a daunting task because of the sheer volume. I would typically receive between one hundred and two hundred incoming project emails every day. Since I obviously didn't have the time to read and respond to every one, my ritual began with first scanning the name of the sender and then the topic. I would read first any message on which I was one of only two or three on the distribution list. Then the urgent items, identified with a big red exclamation mark. Next to be looked at were those that came from people on the front lines; I would pounce on these with some degree of anxiety—especially any from the ATLO team.

While this filtering system was the best I could do, I knew that some important emails would fall through the cracks. I told people, "If it's important and you don't hear back from me, call or come find me."

After the emails, if I wasn't completely inundated with essential meetings, I would walk around and talk to the teams in the working areas. On a typical day, there might be from a few to many comments that translate as, "We don't know where to go from here." Or, "We're stuck, we need to figure this out." I just might have a ready answer, but more often than not I was as stumped as they were.

All of this was just the very beginning of a typical workday. I would usually work until about 7:00 or 7:30—a twelve-hour day. Then it was home for late dinner, followed by another hour or so, mostly dealing with the backlog of emails, finally turning in about 11:30 or 12:00.

After the grind of Pathfinder and Spirit/Opportunity, I had promised myself that I would stay away from work at least one day a week so I wouldn't go bonkers and so that I would have a regular place in my family's life, and they in mine. I tried as well to be home on weekends. On MSL, I usually worked five days a week and on Saturdays worked from home when I could, altogether putting in fifty to seventy hours a week. The problem with flight projects is that they are all-consuming.

I'm not looking for a medal. Most of my teammates were just as manic as I was. I chose the work, I love it, and, when a project is in trouble, many other people at JPL work just as hard as or harder than I do.

* * *

It wasn't a surprise to anyone that writing the software for MSL would be a massive task. While we had been able to get a head start on some items by pilfering software code from the Spirit and Opportunity mission, most of it had to be rewritten, or written from scratch. More challenging, much of the software design and coding had to wait until hardware designs were finalized. By early 2008, the code writing and testing effort had barely gotten under way.

Although the cruise software was pilfered, new software would have to be created for every part of the EDL operation. The software team also needed to create new commands for all the activities connected with doing the roving science mission, such as moving the rover, providing images of the terrain, operating the robotic arm to collect sample material, and so on. This software would have to track positions of each of the arm's joints as well as all the forces acting on each part of the arm.

Other software was being written for guiding the arm for drilling and for transferring soil to the instruments, as well as making sure the rover arm never touched rock or soil with enough force to damage it or damage the rover. The list goes on and on. Indeed, the software would end up being nearly ten times bigger than for Spirit and Opportunity and a hundred times bigger than for Pathfinder.

Creating the software just for controlling the robotic arm was proving to be a massive task. The arm was capable of so many different movements, combined with the need for avoiding situations that could cause damage, that the code just for the arm and its components eventually ran to hundreds of thousands of lines of C code for hundreds of commands like "ARM_ PLACE_TOOL" and "ARM_STOW."

The arm itself would provide a fairly decent imitation of the arm of a live (but huge) geologist, promising to be one of the cornerstones of MSL. But it was a complex piece of hardware. Seven feet (2.1 m) long, made of titanium and aluminum, it was provided with a shoulder for moving the whole arm up and down and side to side, a bending elbow, and a wrist that could turn in either direction as well as move up and down. Because of the way the arm was mounted on the rover, we often included in our descriptions to visitors, "It's a lefty."

Several of the most important tools of the rover would be located on a turret at the end of the arm. In the middle of the turret were the drill and the scoop, and the sieve to filter out larger rock powder brought up by the drill. A device called the "Portioner" would apportion the filtered material to the two instruments for analysis, providing only that baby-aspirin-sized dollop of extra fine powdered rock to either the CheMin or SAM instrument inlet port.

For measuring the elemental composition of nearby rocks, Ralf Gellert's APXS instrument was mounted on the turret at the end of the arm. On the other side of the turret would be Ken Edgett's Microscopic Hand Lens Imager, positioned to take pictures of things as varied as close-ups of the hole that the drill had just created, to views of the landscape, to those space-age "selfies"—Curiosity's photographs of herself.

There were parts of the project where progress was reassuring but at times those seemed to be the exception. A piece of news from the Phoenix lander made our sample-collection team nervous: Their data and images showed the Martian soil in places to have the consistency of thick mud. The Phoenix crew reported they were finding it could be a big challenge to collect powdered samples for processing—imagine ice cream becoming stuck in the scoop, with no finger to dig it out.

On Earth, the higher humidity of the air removes excess electric charge from a sample and provides a natural lubricant for fine soils. On Mars, with dry air and rocks in which water might be chemically bound to the minerals, the granular material from the drill might stick tight.

But then someone—I no longer remember who—asked, "What happens if there's a wind?" We found out from the Phoenix team that the winds on Mars were sometimes blowing away the fine powder in their samples. Putting together a test set up that would blow air—a thin stream of low-density, Mars-like air—while we dropped a sample, the team found that even a light Martian breeze could push the sample away, with none of it dropping into the hole. It would be sort of like that silly episode where the Roadrunner keeps blowing out the coyote's dynamite-stick matches.

This meant retrieving and delivering finely ground bits of rock to the on-board labs for examination would be far more challenging than anyone had anticipated. We might never be able to extract the material from the drill; we might never be able to keep the sample from blowing away. The possibility hung like a big cloud over our whole design. Earlier missions had been dedicated to learning how to land on Mars, and taking baby steps toward learning how to rove and gather scientific information. For MSL, getting samples into the rover's on-board laboratories and sending its scientific data back to Earth was the project's raison d'être.

After all the money and effort lavished on Curiosity, it seemed as if one of the fundamental purposes of the project might be undoable. With only a year to launch, I wondered if the combined brains of the scientists teamed with our engineers and technicians would have time come up with solutions.

This work would be much easier if all of the really challenging problems could be addressed early on and you would eventually reach a point, maybe a year, or a year and a half, or two years before launch, when the problems that popped up were all relatively minor—nothing that couldn't be solved in a month or two at the most. Well, when we signed up for MSL, nobody promised it was going to be a piece of cake.

Even the EDL effort and sky crane, which had been moving along relatively trouble-free until then, turned up with unexpected problems. First we found that when the rover wheels popped free as the rover was descending on the bridle ropes during the sky crane maneuver, they popped out so fast that some of the hardware was in danger of breaking. Then in late 2008, we decided to run a test scenario to look at the moment at the start of the sky crane phase when the rover separates from the descent stage to become suspended for landing. Until then, it had been getting power from the descent stage, but at that moment the source of the rover's electrical power changes. The electrical cables that provide power between the descent stage and the rover are cut, with the rover shifting to its own internal battery as its power source.

When we simulated the electrical effects of cable cutting, for some reason the descent stage and rover briefly stopped talking to each other at what would be a critical moment during the landing. That was very odd. John Wirth, our lead ATLO electrical test engineer, dug out a few old-fashioned-looking oscilloscopes that show waveforms of voltage and current moment by moment so that we could examine the signals.

It didn't look good. Victor Moreno, our lead electrical systems engineer, joined me to study the plots. It appeared that after the separation of the rover and the descent stage, short-circuits associated with the cable-cutter operation were changing the power supply voltage balance. This damaged some of the circuits used to send messages between the rover and the descent stage. Based on what we discovered, we put together a Tiger Team to analyze the situation and figure out a solution.

It turned out to be one of those situations where a little miscommunication causes a gigantic problem. The designers who picked the type of circuit used to send messages between the rover and the descent stage had chosen a type that is extremely sensitive to changes in the power supply voltage. They weren't aware that this circuit would need to travel up the long cable from the rover to the descent stage.

I looked at the schedule. I could not see any possible way we would have time to redesign the circuits. To fix this, the ATLO team would have to take the rover apart once again, pull out a lot of the computer network wiring, isolate the connections between the rover and the descent stage, and make extensive changes to parts of the electronics and the cabling. It was a little like trying to make a beautiful, elaborate chest of drawers as a Christmas present for your wife—and starting on December 23. By my reckoning, the work of isolating the connections and making the other changes couldn't be in place until a few months before the spacecraft had to be packed up and ready to ship to Cape Canaveral for launch.

One or another of the younger team members was forever asking, "Is it always like this?" I would try to reassure them: "No, this is the toughest project I've ever seen." In less than a year, I had gone from incredibly enthusiastic to incredibly depressed, a 180-degree flip I had never experienced before.

Despite the numerous problems we had discovered so far, in principle all of them could be fixed in the little time we had left before launch. Management was still trying to be bullish about meeting the 2009 launch date. There was little else that they could do but stick with a "can-do" attitude.

I still had not found any metric to satisfy the challenge Gentry had presented me with in the mall eight long months earlier. And the problems on MSL weren't the only frustrations in my life. Anyone doing as much walking as I was doing in a typical workweek should have lost weight. I was gaining. I knew this was in part the result of eating too fast, out of urgency, anxiety, and frustration. We all were struggling with learning how to triage the mounting problems, and found ourselves only having enough time to deal with the urgent ones. Problems that were merely important but not urgent simply fell by the wayside.

I wanted to tell people, "This is not how it's supposed to go." For some problems I sorely wished I could say "Take it off the spacecraft, send it back to the lab, and make sure it's fixed before it's returned to us." But I knew the only sensible reply would be: "Rob, we can't do that. It would just put us even further behind. We'll have to live with it as is."

Instead we plowed forward, feeling as if we were drowning.

At our monthly review in late November 2008, all the top project management and some of JPL's upper management were in attendance. As usual, each of the subsystem delivery managers stood up one after the other and gave his or her status.

Then ATLO manager Dave Gruel and spacecraft manager Matt Wallace stood up to talk. Dave said, "We now no longer have enough schedule margin to meet JPL's margin policies."

JPL policy requires that a project must have at least 15 percent of its scheduled ATLO time unallocated. The ATLO schedule is laid out well in advance—what hours on what days each team will be able to run its tests or bring in new hardware to integrate. But the hardware had not been arriving on time. When a piece of hardware did arrive, Dave had to give the team some of the unallocated time, and what he was saying was that he no longer had the required 15 percent of unallocated time. In fact he was nearly negative. There was no slack left.

Project manager Richard Cook asked, "Dave, what about weekends?"

"They're all scheduled."

"How about second shifts?"

"All scheduled."

It was as if a cold chill had suddenly descended on the entire room. Matt in particular was dejected. He perhaps more than anyone had never given up hope; he had pushed and pushed, determined to get MSL to the launch pad in 2009.

My own reaction was different. This was my missing metric, the one I had been so desperately seeking—a hard fact that proved we could not launch on schedule. I was relieved that we finally had hard data everyone would accept. Though I was of course sad that we had failed to make our schedule and still worried that NASA headquarters might want to pull the plug on MSL, nevertheless a great weight had been lifted from my shoulders. If we were given a 2011 launch, we would have the time we so desperately needed to solve the many problems vexing us.

How many times in my life had I had a meeting that was at the same time both depressing and relieving? I couldn't think of any. This was a first for me in too many ways.

No one wanted to mention that NASA headquarters had only two choices: Tell us our launch date was being slipped to 2011, or give us the ax and spend the rest of the MSL budget on other projects.

Instead there was a discussion about who was going to present the bad news to headquarters. The decision was that Richard and Tom Gavin would go to see Dr. Elachi. I found out later that on hearing the news, he put in a phone call to NASA associate administrator Ed Weiler, who shared the news with NASA's top man, our onetime champion, Mike Griffin.

Elachi must have made that phone call knowing that no previous Mars mission had ever progressed so far yet missed its launch date. Postponing our launch for two years would be costly. It would push NASA's expenditures for MSL about 30 percent over the amount originally allocated by NASA and Congress. When that happens for any federal agency, the agency must report the facts to Congress for a decision of "Quit now" or "It's okay to proceed."

Gentry Lee likes to tell a story from the very early 1970s, when the Viking project was in a similar situation and, as with MSL, needed congressional approval for the additional funding to postpone the launch for two years. The news brought a heated reaction from a Texas senator.

"Two years late?"

"Yes, sir, we have to wait two years. We can't launch any sooner than that."

"Dammit, when I want to go to Abilene, Texas, I get on the bus and go to Abilene, Texas!"

"Yes, sir," was the response. "Well, the bus to Mars only leaves every twenty-six months."

In addition to that Viking project, missing a launch was actually nothing new. The Shuttle and the Space Station were vastly late and vastly over budget. First-of-a-kind projects are nail-biters all the way, for reasons that you'll find many examples of in these pages. Many have gone way over budget, others stayed within budget but failed.

At the same time, it wasn't hard to understand the dilemma NASA would be faced with. Among other projects in the queue for major funding was the huge James Webb Space Telescope (JWST). Intended to replace the Hubble Space Telescope, JWST has been described by NASA as providing a way of studying "every phase in the history of our Universe, ranging from the first luminous glows after the Big Bang, to the formation of solar systems capable of supporting life on planets like Earth, to the evolution of our own Solar System." The value of the knowledge that JWST could provide is incalculable. Yet the cost of delaying MSL would almost certainly mean dipping into funds that would otherwise be used for moving ahead with the Webb and other worthwhile projects.

One element I hadn't counted on but hoped would make a difference: NASA and much of the planetary science community had come to believe that even with a higher cost than originally anticipated, MSL's scientific promise was beginning to look huge. But on the other side of the fence, a lot of scientists who didn't get instruments on MSL were upset at the idea that a postponement for us meant we would be gobbling up what could be the budget for the next potential mission, and another opportunity for left-out scientists.

When authorization came from Capitol Hill in February 2009, JPL received the official decision from NASA headquarters, but it came as a pair of emotion-stirring pronouncements. The good news was that we were being given an additional two years: NASA was slipping the launch date to 2011. To cover the costs that this delay would impose, it was increasing our budget by $400 million. The bad news was that NASA headquarters was pulling the plug—cutting off as much of MSL spending in fiscal year 2009 as it could so that other projects would be able to move ahead. The bulk of our spending authority would be returned in 2010.

We were to put the rover into mothballs—actually wrap it up and put it in a corner until authorized to go back to work on it. In effect we were being told, "You're only a year behind. Do only small essentials now, and we'll let you start spending again when you're about one year from launch."

Apparently we left them with the impression we were just running a little behind, and they were now giving us an extra year. Hmmm. Sitting in Washington, I could see how that appeared to make sense. After all, in the ATLO High Bay sat what looked like a fully functional rover, already being tested. Sitting in Pasadena, what we were facing was a much different picture. We were asked to downscale our workforce to one-third the size. With fewer people and a huge list of unsolved problems, I knew we would be facing a manic time trying to find solutions. Many in the huge team we had trained would be leaving to work on other projects.

Over the next two weeks, I hurriedly tried to interview nearly everyone on the project, asking what technical problems they had been facing. I added their new items to the list I had begun at the offsite, my "retreat" months earlier.

This list had acquired a nickname: The MSL gang started calling it "The Manning List." Perhaps they considered it a dark badge of honor to have an issue on The List. For me it became an albatross, to quote Coleridge, that about my neck was hung.

CHAPTER 14
Shutdown and Restart

At a press conference announcing that Mars Science Laboratory would miss its planned launch date, Mike Griffin cited "lingering technical problems" and "a backlog of unresolved work and undiagnosed problems with the rover's actuators, motor-driven gears that move the spacecraft's wheels, bend its robotic arm, and drive its drill." He also said, "We've determined that trying for '09 would require us to assume too much risk—more than I think is appropriate for a flagship mission like Mars Science Laboratory." Said Doug McCuistion, director of the Mars Exploration Program at NASA Headquarters in Washington, "We've reached the point where we cannot condense the schedule further without compromising vital testing."

I had to agree. To my eyes, the reporting was pretty accurate, even though we had a lot more challenges on our plate than most people knew. An article in *New Scientist* magazine reported, "NASA will postpone the launch of its over-budget Mars Science Laboratory (MSL) rover by two years, to 2011. The delay will add another $400 million to the cost of the mission and will probably force the delay of other agency missions, officials say." The article noted, "The added delay will bring the total lifetime cost of the rover mission to more than $2.2 billion. MSL is already $300 million over its proposed 2006 budget. . . . NASA may have to draw money from other Mars missions as well as the agency's larger planetary exploration programme to pay for the MSL delays."

The magazine piece went on to mention a recent op-ed in the *New York Times* by our onetime nemesis Alan Stern, in which he criticized such overspending, "arguing that they sharply limit the number and capability of missions the agency can undertake." True enough, but that doesn't take into account the great many scientists for whom no space project held more promise than MSL.

As if rebuking Stern, Mike Griffin told the reporters, "We know how to control cost—just build more of what you built the last time." In other words, the work is easy if you just keep doing what you already know how to do.

The *New York Times* story gave the reasons for the postponement as "lengthening delays and lingering technical issues." Well, they were "technical issues," and they certainly were also "lingering," but that description didn't quite match what we on the MSL front lines were seeing and experiencing. We had our work cut out for us.

After the holidays, we returned to cope with NASA headquarters' plan to take much of our 2009 budget. Armed with the Manning List, we made a strong case in favor of our being allowed to find design solutions for these problems before the mothballing. We also argued that we had made it this far and put together most of the spacecraft, albeit with missing pieces, and we should complete some of our system tests as a way to reduce the risk of finding new problems later. In particular, we wanted to confirm that the full-up spacecraft would work in the cold vacuum of space.

Richard Cook persevered and NASA headquarters gave in, allowing us to keep a decent-sized engineering staff to solve some of the most urgent problems and complete the important tests on the flight vehicle before mothballing it, plus enough of a small additional workforce figuring out solutions to other problems, ready to put the solutions into action when the all of the flight equipment was back in ATLO.

We didn't know whether to celebrate or go into hiding. Two more years at the same exhausting pace was hard to contemplate—for us and perhaps even more for our families. Some core team members couldn't face the prospect, and announced they would be leaving. I couldn't blame them. For those who stayed, the thought of having to keep this pace until landing in 2012 and beyond must have been daunting. Some old faces returned. Pete Theisinger rejoined MSL's Project Management team and Dara Sabahi and Joel Krajewski joined to help organize the systems engineering team.

With help from Dara and Joel, we quickly teamed up to prioritize the Manning List and get action started. We created seven "Red Teams," one each for actuators, avionics, electrical, power converters, sample-handling, test infrastructure, and fault protection. Each of these teams would be handed issues to close out from the Manning List. Once a solution was found, they would estimate how expensive it would be to implement; project management would decide if they considered the fix worth the cost.

With a new plan and a few people to help, we raced off to try to solve the many frustrating problems we faced.

For the sample-handling team, two major aspects hadn't been worked out yet: the sticky-soil problem and the wind problem.

For sticky material, we already had shakers and thwackers built into the CHIMRA device on the end of the arm. But we also needed some sort of tools to be able to clean out material that remained stuck. Daniel Limonadi's sample handling team came up with a series of "widgets" that could be mounted on the front of the rover. Some of these could be used as cleaning tools to scrub out tight spots on the end of the arm that contained the filters and the portioners. The arm could be positioned to clean plugged areas on an object more or less fashioned after a pipe cleaner.

They also added a "sample playground" to the front of the rover. If the rover drilled into rock that appeared too sticky, a bit of the material could be dropped off into a sample tray, where the scientists could study its composition before moving it to filtering and portioning. If the material got stuck in the drill itself, we had already made provisions for two spare bits that would replace a permanently plugged drill.

To keep wind from blowing away a sample, the team came up with wind guards that surrounded the inlet doors. Testing convinced us that we could now handle wind and most clogging.

We also learned that some of the rocks on Mars might have "deliquescence." This is a property of minerals that have a lot of H_2O molecules as a component. When you drill into a rock with deliquescence, the powder from this dry rock can suddenly turned into the consistency of peanut butter. We were never able to satisfactorily solve this problem. Even our best cleaning tricks couldn't deal with pasty rock powder. Our only workaround would be to try to recognize that the rock a scientist wanted to drill had deliquescence. A plan was worked out so that before drilling into a new type of rock, a tiny divot would first be drilled; the material retrieved would be imaged with the MAHLI camera and the images sent back to Earth, where they would be examined to determine if it was okay to proceed. Some problems have no hard and fast solutions, and deliquescence is certainly one of them.

We had also been trying for months to solve the "not enough power" problem. Although we made changes to minimize the number of electronics items that had to be powered, especially during the long sleep hours, it was clear that that these changes would not be enough.

This forced us to make a difficult concession: we would save power by running only one "string" of electronics at a time. We had planned to keep both computers and much of the supporting electronics running, so that if the main, "A-side" computer failed, the "B-side" backup could immediately take over. When the rover was asleep and the computers were turned off, a pair of small "lizard-like" brains—a main unit and a redundant unit—would monitor the rover (and themselves), immediately awakening the computers in an emergency.

With the backup computer always turned off, this meant that in case of a failure of the main computer during the day, we would need some way to independently detect the failure and turn on the backup. These changes required a radical redesign of the rover's fault protection, requiring that a part of the rover's internal cables be rerouted and a huge part of the software be reworked.

The new design would allow the nuclear power source to recharge the battery at night, while all but the absolutely essential items were shut down. But each morning, just drawing enough energy to warm up the robotic arm—that one chore alone, necessary before the useful work of the day could get started—was enough to render the battery nearly empty and leave the rover asleep and useless for the rest of the day.

Now that we had time to fix the problem, we were faced with a painful but unavoidable decision: The battery would have to be replaced with another battery of twice the capacity.

In early 2009, I sat down with Dara, systems engineer Chris Salvo, and Peter Illsley, the mechanical lead design engineer for the rover, who had proven repeatedly to be a can-do guy. Our message was, "We have a big problem and we need your help. We have to double the battery size."

It looked as if Peter's head was going to explode. Clearly, he thought we were asking for the impossible—yet his history suggested he could come up with an answer.

Several weeks went by. One day Peter and Chris dropped by the MSL trailers to say, "This is what we think we can do." They had found just enough room to be able to reorient and then stack a pair of batteries with a total volume about twice that of the original. But Chris made it clear: "Someone will have to design a new battery that will fit."

Okay, this wasn't going to be cheap. What they were suggesting would involve going to the battery company and asking them to design a pair of batteries that would fit in the available space. They would have to be custom-built. Fortunately, the battery that the Phoenix lander used was close to what we needed, though ours would have to be 36 percent bigger. Our battery manufacturer, Yardney Technical Products, in Rhode Island, agreed to design and build the new batteries for us. But with a catch: It would take over a year.

On our end, making room for the new battery would mean an enormous amount of undoing and redoing. We had to partly disassemble the rover so we could pull the old battery and support structure. Working out a complex three-dimensional puzzle, the rover mechanical team had to figure out a way of getting the new battery in without excessively rearranging the many components. The effort of providing enough space for installing a battery twice as large as the original one took hundreds and hundreds of man-hours; the total cost was over $2 million.

On the plus side, it would essentially provide a solution to most of the power problems that had haunted us for two years.

As the time approached for putting the rover into "mothballs," we tested the whole spacecraft together—rover, descent stage, aeroshell, and cruise stage—in the large thermal vacuum chamber at JPL. After that, it took almost two months of intensive work for full disassembly. The crews removed the electronics units, to send back to their developers. They would have the rest of 2009 and the first half of 2010 to test the engineering model units we had been using, make all needed improvements and subject them to rigorous testing, then deliver flight units back to us, ready to be reinstalled.

In the High Bay 1 clean room, the crews wrapped the cruise stage and aeroshell in plastic, sealed them securely, and moved them out of the way to await 2010 and the restart of the money flow from NASA headquarters. Meanwhile, rework began on the descent stage to fix problems we had found in the descent propulsion design, while the cable harness team went to work modifying the rover's electrical cabling for the new redundancy and fault protection design. After that, the rover, too, was put away securely. Much of our ATLO test team was sent to other jobs, waiting for the day in mid-2010 when they could start assembling and testing again. When I thought about how far we still were from a ready-to-launch spacecraft, it only made me more uncomfortable, but at least we were on a productive path that had time to succeed.

For months, everything had been moving so fast that there had been no time to document the changes being made. Even with the restraints of our slashed budget and reduced workforce, Dara and the project management understood that our culture had to change from the mad rush of people stumbling over each other, to slowing down, with decisions being made in linear fashion, one at a time, and accurate records and rationales for each change being kept. This was certainly the right thing to do, though at the same time it scared me that the documenting and approval process might take on a life of its own and slow us down.

By late 2009, the actuators—the motors and their gearing—finally arrived, and the teams set about testing them individually before installing some of them on the rover test bed and delivering the flight units to the arm and other mechanical teams for installation into the mechanisms. Finally, drilling tests using a real arm and real motion on real Earth rocks could begin in earnest.

Testing a drill sounds as if it should be pretty easy, but testing a Mars rock drill is another matter. To begin with, we don't have many rocks from Mars to drill into. There are some one hundred very rare Mars meteors that have been found around the world, but these are too precious to test with. Instead we settled for Earth rocks that were "analogs" to the kinds of rocks the rover might encounter. The drilling would take place in a cold vacuum chamber that would reproduce conditions on Mars. Bob Anderson of JPL, a rare combination of geologist and engineer, rounded up rocks for Daniel's team to drill into. Some rocks drilled like butter, while others mysteriously caused the drilling progress to come to a halt. The team would have to tune and retune the combinations of percussion strength with the force, angle, weight, and rate of drilling. Project scientist Grotzinger's comment about this was, "We sweated a lot of details."

Ultimately, the final drill bit design was tested by subjecting many identical non-flight bits to boring 1,200 holes on dozens of types of rocks. Of course, the flight drill bits would be virgin; we would save their first drill for the real thing on Mars. Three bits were precision cleaned and sterilized, then put on the rover—one on the end of the arm, the other two as spares, attached to the front of the rover. To avoid the danger of dulling them, the flight bits were never used to drill rocks on Earth.

Unable to test EDL's sky crane landing phase as a whole, we began testing the individual elements separately. For the throttled rocket engines, this meant simply putting them on test stands and running them one at a time. For the radar that would feed data on the distance and speed to the on-board computer during the approach to the Martian surface, there were two tests. For one, a radar unit was fastened inside the nosecone of an Air Force F-18 fighter jet; the pilot climbed to miles above the desert floor, then nosed over and dove straight toward the ground before pulling up at the last minute.

In the other test, the radar was attached to the front of a helicopter with a mockup of the rover hanging below as the helicopter lowered the mockup slowly toward the ground while we watched the radar data to verify that it was accurately measuring the distance to the surface. (I'd love to have been offered a helicopter ride, but no volunteer copilots from the team were permitted.)

For a look at the performance of the rover separating from the descent stage, we lifted the engineering model of the descent stage and rover to the ceiling of one of our test buildings, and then, while we collectively held our breath, initiated the sequence of pyrotechnic explosions that cut the cables and released the bolts that held the rover and the descent stage together. Like a climber being lowered down a rope, the bridle umbilical device paid out its three bridle ropes to swiftly lower the rover, while the pyros on the rover simultaneously fired, releasing the rover wheels from their stowed position. The rover's motion stopped just short of the floor when the ropes had reached their full length. We were recording all the data, of course, but our eyes gave us the essential answers: Yes, the wheels of the rover do drop into place just as intended. Yes, the bridle umbilical device that pays out the ropes and the electrical cable worked as designed. Yes, the rover does withstand the dynamics without falling apart or experiencing stress or damage. (You can see a video of this online at CuriosityRover.info/DroptestVideo.)

Meanwhile, we continued testing the whole EDL system in a simulated Mars world millions of times using a fantastically detailed computer simulation. For each simulation, one or another or several of the many parameters would be given a different value. The parameters included items such as density of the atmosphere, strength of the winds, surface characteristics of the terrain, drag of the parachute, strength of the thrusters, friction of the cables, and many others. There ended up being thousands of them. If we didn't know precisely what to put in for a parameter, we made sure that the simulations tried out the whole range of possibilities.

We were to discover after Curiosity had landed on Mars that we had missed a crucial item. The long list of variable parameters had not included one that should been obvious: gravity. In the simulations, the EDL team used a fixed value for gravity that was rather generic for that part of Mars. We failed to take into account that the shape of the surrounding terrain and hills might affect the actual gravity, and because we didn't try other values, we didn't notice just how sensitive the landing was to being slightly off with the value the team had chosen. The value for Mars gravity used in the simulation turned out to be slightly too high—*very* slightly, only 0.1 percent—but significant enough that Curiosity's slowest-ever landing was even slower than we expected. If EDL had taken much longer, the lander could have run out of fuel, the rockets would have been unable to slow the craft sufficiently, and the landing would have turned into a disaster.

If instead the mistake had been made in the other direction, the consequences might have been just as bad. In this case, as well, the rover would have hit the ground too fast and too hard.

Just as with the Mars Climate orbiter navigation mistake, this was a mistake no JPL team will ever make again.

As is always true in life, the more pressure we are under, the more welcome a good laugh is to break the tension. We didn't find anywhere near enough occasions to laugh, but I vividly remember one example. The professorial, ultrasmart Miguel San Martin often loads his sentences with acronyms. In one late review-board session, he proudly told the group, "The DMCA, running the MLE's PD rate controller off of the IMU, is robust to CG offsets after the CLF command from the RCE." After a moment of dead silence, everybody in the room broke out in a roar of laughter.

Miguel scanned the faces, bewildered, not understanding what all the laughter was about. As incomprehensible as that statement seemed, we had all, to our great surprise, discovered that we understood him.

At that point in the daunting race to finish on time, a hearty laugh was greatly welcome.

By the end of 2009 we had fixed, or at least had plans for fixing, many of the most significant Manning List problems, nearly completing the work we had set out to do during this challenging period. We had more than the usual reasons to celebrate the new year: soon our funding would ramp back up, the ATLO team would return in their bunny suits, the spacecraft sections, with their plastic wrappings removed, and the rover restored to its place near the center of High Bay 1, and we could move ahead with the work of getting Curiosity and the rest of the spacecraft ready to launch.

In 2010, the project was lucky to be able to get back many of the designers and engineers who had left the previous year. I was worried that a horde of them would be so tired of the MSL slog that they wouldn't want to come back. We welcomed them like the return of lost soul mates. Apparently, they loved the attention we lavished on them. (Jim Donaldson pointed out that, "When you're in our line of work, sometimes you confuse attention for love.") They now began debugging and figuring out how to implement the many changes we had made and how to work around the many design defects from our haste to build up MSL in 2008. Fortunately, we were able to designate many of these items as "UAI PFRs" (standing for "use-as-is PFRs")—meaning they were defects we could live with and we were accepting without fixing.

We were barely able to do nominal activities without stumbling over a weakness in the design. Gradually the software was modified to work around these defects so that instead they became "idiosyncrasies" rather than outright problems. Still, Curiosity and the whole of MSL remained more fragile than it looked, especially in some of the avionics, grounding, software, and interface designs. As we dived into the details using the test beds, we continued to find more "idiosyncrasies" that had to be dealt with.

The money spigot was turned back on slowly at the beginning of 2010, and with the spacecraft unwrapped, tested hardware began to return to High Bay 1. Naturally enough, project managers Pete Theisinger and Richard Cook didn't want to see money spent on anything they weren't convinced was absolutely necessary for the project. When Jim Donaldson's avionics team (to use just one example) kept finding dozens of problems, Jim would take his list of problems and try to get permission from project management to fix them. The response from Pete and Richard was, "We don't want to spend time and money to fix these problems unless you can prove that we can't live with them"—meaning MSL might fail if the items weren't fixed.

For example, we were using a lot of first-of-a-kind, state-of-the art, super-high-density programmable chips that contained an incredible amount of circuitry, shrunk to fit inside a device the size of a quarter. Unfortunately, these expensive chips were not reprogrammable. If a redesign or a bug fix required replacing a chip—which at times happened as often as twice a month—the cost of each replacement chip ran around $30,000. Testing each replacement version would require the effort of two to four people using a whole laboratory of equipment and would typically take many weeks. Meanwhile, the test bed team would have to wait for those tests to be finished, and would then have to repeat many of the tests they had done with the old versions of the chips. If you include lost time plus the salaries for the work hours that the teams spent retesting, this added another $20,000 or so to the cost of replacing a single chip. So it wasn't hard to understand that management was treading carefully on any request likely to involve that degree of cost and labor.

If it looked as if there was a way to work around a chip problem using software, Jim and the software team would reluctantly spend months looking for a solution. Some of these workarounds became clumsy Band-Aids that also had bugs and in a number of cases would haunt us until well after the rover had landed.

By June, the rover's innards had finally been assembled. Still, the frequent "No's" to requested fixes that Jim felt would make his avionics subsystem less buggy led to tension and frustration. Not surprisingly, from the perspective of the subsystem engineers like Jim, every "No" increased the complexity of MSL and possibly increased the risk of some sort of failure or anomaly, even if that failure didn't kill the rover. The fear of failure is something that gnaws in your gut. In this line of work, that feeling never completely goes away even after the spacecraft has landed safely.

On a personal level, my physical condition was now significantly improved. A year or so earlier, my doctor had found that my blood pressure had gone through the roof, and was much concerned about my hypertension, and I had added some fifteen pounds (6.8 kg). Based on that dire report, I had started carving out time to go to the gym, spending two to three hours a week there. The workouts were helping me get through the stress of the job. At the same time, I felt a little happier with the way we were getting things done, solving problems one by one by one, paring down the list. I was feeling better from a sense that the project was getting back on track. Still, the length of my list of problems not yet solved was enough to keep me scared part of every day. The work hours and pressures were as intense as ever. I knew it wasn't going to get easier any time soon.

Finally, the payload of scientific instruments was installed, and the cameras were put in place. I imagined I was seeing more spring in the steps of the team members as they started to feel that the system was coming to life—even though it was still very buggy, with a ton of small items that remained to be fixed.

In June 2010, we at last had a fully built rover and began testing whether the hundreds of thousands of assemblies would work together and whether they would receive the correct commands from the millions of lines of still mostly untested software code.

One of the special areas of concern was whether we had made the rover sufficiently *self-reliant*. Buried in all of that software code, there needed to be

FIGURE 13. The fully built rover. The rover is 9.5 feet (2.9 m) long and about 9 feet (2.7 m) wide, and the top of the mast is about 7 feet (2.2 m) above the surface. A time-lapse video of the construction is online at CuriosityRover.info. (Courtesy of NASA and JPL/Caltech)

instructions for detecting when something went wrong, and instructions on how to correct the problem or work around it. Anything that could take the rover down—short of a major catastrophe like losing a wheel—needed to be fixable. If a radio died, for example, the software must be able to detect that and switch operations to another radio.

Meanwhile, the ATLO team was not overly fond of the fault protection efforts. They much preferred to be in charge of the rover themselves. With fault protection running, on occasion the rover would suddenly assess that something had gone wrong; it would turn off equipment and turn on other equipment. Of course all this would happen without the rover asking permission from the ATLO test conductors.

We were regularly getting urgent midnight phone calls complaining that the rover had taken over and that the ATLO test was ruined. Telling them that they should be nicer to the rover did not seem to help. The real problem was that in ATLO, the spacecraft was not being and could not be safely operated as it would be during the mission, with such different activities and stresses during launch, in deep space, during landing, and on the surface of Mars. Eventually we reluctantly agreed to disable the fault protection during testing, except for essential tests that could not be run in any other way.

On July 22, 2010, the day came at last for the first modest test drive of Curiosity—with all of us feeling something like the anticipation of young parents when their firstborn is about to take his or her first steps. The first drive would be a short one. The crew laid out a special clean blue matting some 15 feet (4.5 m) long, soft and electrically conductive to avoid sparks. We watched as the rover ploddingly made its way about five feet before turning in place and going back to its starting position. That was enough for now.

A few months later, in early September, we finally got a chance to put the rover through its paces. This time about 40 feet (12 m) of the matting was laid out. Metal triangles were positioned as speed bumps that the wheels would have to drive over, simulating the way the wheels would have to handle rocks along the pathway on Mars.

This wasn't exactly the Indianapolis Speedway: Curiosity was going at a snail's pace of about 0.09 miles per hour (4 cm/s), its top speed. On Mars, when using the software and its engineering cameras to watch for and ponder obstacles, the rover would progress at about a tenth of that.

Overlooking the ATLO floor, the balcony was jammed with a large crowd of team members watching anxiously and snapping pictures with everything from cell phones to serious, professional-quality cameras. As soon as Curiosity had made its first visible movement, the crowd let loose an eardrum-pounding cheer.

FIGURE 14. Curiosity on its second and final—and most demanding—test drive. In the lab, the engineering team had it negotiate "speed bumps" in order to evaluate its ability to travel over uneven Martian terrain. (Courtesy of NASA, JPL/Caltech)

To me it was a wonderfully symbolic moment. It was reassuring to see the rover operating as a completed system. Finally. Still, there were some aspects to the drive I was quite nervous about. The software that was controlling it was—even this late in the game—still a preliminary version, so we weren't sure yet that the code would do a good job monitoring that the motors were not overheating. Anxious over whether the software was doing its job properly, I kept racing into the ATLO control room and peering over the shoulder of the test engineer to make sure that the data flowing from the rover was telling us everything inside it was still normal.

Some twenty minutes crept by before the rover reached the first speed bump. A front wheel slowly rose about a foot over the top of the bump, and slowly down the other side without a problem. These two tests were just enough to give us confidence that the rover was built properly and that the equipment worked. These were the only drives Curiosity would make before landing on Mars, with most of the testing being done with the test bed rover.

For all of us, this was an exciting milestone. Finally we had a sense that we were really going to get there.

CHAPTER 15
The Final Stretch

For years, the engineers and scientists involved in Mars projects had been at odds over the selection of landing sites. The places that the scientists identified as most likely to show signs of water and the ingredients necessary for life were invariably places the engineers identified as impossible for landing safely. The scientists wanted to land at places like the equivalent of our Grand Canyon, while the engineers like me insisted on landing at places more like Oklahoma, with flat landing surfaces and few boulders, offering little danger of the rover's tipping over or falling off a cliff at touchdown.

On previous Mars missions like Pathfinder and the Spirit/Opportunity twins, I had to face the frustration of continually telling the science team that the scientifically compelling landing sites they wanted were not places the spacecraft could safely land. On all three of those missions, this cat-and-mouse game happened repeatedly until finally the scientists would identify a place where we could have some confidence of providing a safe landing.

This time, with the new entry, descent, and landing concept, the scientists would have much more liberty in selecting the landing site than anyone could have imagined back when MSL was just a dream. More than fifty sites had been on the initial list of possible landing places. Mike Watkins, MSL's mission manager, adeptly organized and hosted five open workshops, with a large number of scientists from all over the world joining the debate. The early criteria included sites with these characteristics:

- Within plus or minus 30° of the Martian equator so as to stay away from climates that are too cold.
- Elevation less than 0.6 miles (1 km) above the mean elevation—sufficiently low enough that the spacecraft would have traveled through enough of the Martian atmosphere to slow it during EDL.
- Most slopes less than 15 degrees.
- No large quantities of rocks bigger than 1.5 feet (0.5 m).
- No strong winds that might excessively cool the rover during the nighttime.
- Few escarpments or cliffs that could trick the radar.
- A surface that is radar-reflective (for obvious reasons), load-bearing (no sinking into sand dunes), and "trafficable" (meaning safe for landing and roving).

Each of the proposed sites was evaluated using data from the Mars orbiter spacecraft. This data, which included rock coverage, terrain shape, and other attributes, was handed to the EDL team; they ran thousands of simulated landings at each site and counted the number of times that the landing exceeded one or another of the system specifications. To the surprise of many of us, with our new robust landing architecture, there were relatively few landing sites that actually failed. Only the bottom of Melas Chasm, one of the more spectacular canyon features on Mars with its brutal winds, was rejected outright.

The scientists were sometimes "very aggressive" in arguing for their preference, and a few of the conversations became acrimonious. In a process that involved some four hundred scientists over a period of more than five years, they batted around the pros and cons of fifty candidate sites.

Late in the process, when the number of sites had been pared down to seven, teams were assigned to argue for each. The final gathering in a series of landing-site workshops was held in September 2008. The teams debated the candidate sites, narrowing down to four finalists—Eberswalde, Gale, Holden, and Mawrth. These four were evaluated using the most up to date computer simulations of entry, descent, and landing. Thousands of simulated flights and landings to each site gave the EDL team confidence that none of these were particularly dangerous. In fact, all four showed very high probability of success.. For me this was a huge milestone. For the first time it wasn't the spacecraft engineers saying, "No, we can't land there." Now the science community could argue the pros and cons for each site strictly on their scientific merit. I began to feel that taking this huge gamble to change the very architecture of how we landed on Mars was already paying off.

Curiosity's landing area was going to be so much smaller than for any previous mission that the scientists were bound to be pleased. In fact, none of the proposed sites would have been possible to land on without our new guided-entry tactics. The smaller landing area meant Curiosity would not have to spend much of its lifespan just trekking across the landscape to its intended location or having to contend with hills too steep or areas strewn with large boulders blocking the path. If all went according to plan, our rover would be landing much more accurately tucked into the selected area, within a landing ellipse only 12 miles by 4 miles (20 km × 7 km), about one-fifth the size of the landing area dimensions for any previous Mars lander.

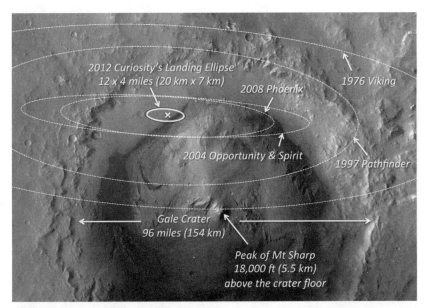

FIGURE 15. The landing ellipses here show the accuracy of Curiosity's guidance as compared to that of earlier Mars missions. Without the use of autonomous entry guidance, MSL would not have been able to land reliably inside Gale Crater. (Courtesy of NASA and JPL/Caltech)

That was the good news. On the downside, Curiosity might still have to drive about 12 miles (20 km) to the target of greatest interest, the base of Mount Sharp. At the rover's snail's pace across the Martian surface, the journey would take about a year.

When the voting by the scientists at the final landing workshop ended in a tie, John Grotzinger turned the decision over to MSL's project science group: the principal investigators and builders of the rover instruments. It was their decision that ruled. John says that, finally, "Like morning mist rising in Kentucky, there was a rising enthusiasm for Gale."

After years of advocating for Eberswalde, even John had found that Gale Crater was the most compelling site. He was personally pleased with the choice but concerned that NASA headquarters would balk, since all the available information suggested that any important scientific revelations would have to wait until Curiosity could plod its way across the twelve-mile stretch to reach the base of Mount Sharp. One statement from NASA headquarters contained a hint of displeasure, suggesting that if Curiosity were to die while trying to get to Mount Sharp, "We will have nothing to show, which would mean a huge waste of taxpayer dollars." The facts would turn out differently.

John enrolled a team to study the proposed Gale site, and they came up with the evidence he was hoping for: Photographs from the Mars Reconnaissance

FIGURE 16. John Grotzinger (standing at front), speaking before an independent review panel and MSL project management team, makes the case for why Gale Crater was the best and safest choice. (Courtesy of NASA and JPL/Caltech)

Orbiter spacecraft showed evidence of rock outcroppings and signs of past water flow right inside the landing ellipse. Even without leaving the landing ellipse, Curiosity would be able to do great science.

That clinched the decision: MSL would land at Gale Crater.

Gale had won the competition for a series of compelling reasons. It covers a wide area, about the size of Connecticut and Rhode Island combined. It lies at a low elevation, so that water could have flowed in and become trapped. More or less in the middle of the crater, towering higher than Mount Rainier, Mount Sharp is a most enticing place to have Curiosity visit. Just as at Grand Canyon you can see layers from different eras of our planet's development, Mount Sharp is clearly revealed in photos from our Mars obiters as offering an entire series of layers that are like a geologist's history book of Mars through many eons. The lowest layers might in fact be ancient enough that they were laid at a time when scientists suspect Mars climate was warm, wet, and mild. Data from the Mars Reconnaissance Orbiter revealed large layers of water-altered rocks—sulfates and clays—just at the foot of the mountain and within reach of Curiosity. One planetary geologist, Brown University's John Mustard, commented, "We believe that at Gale Crater, we have located that boundary where life may have sprung up and where it may have been extinguished. That's why we're going there."

In January, I found that it would be nearly impossible to fully test the entry, descent, and landing software and sequence of events on the real vehicle in ATLO. The problem: hardware safety issues. For example, we couldn't turn on the radar in the High Bay for long without a danger of the radar overheating, and we couldn't run the eight descent engine throttle motors long without risk of degrading them.

What's more, although the rover transmits data during EDL in real time, it is still too little to be able to monitor everything that the ATLO team needed in order to safely monitor the hardware during these tests. EDL would have to be run in a nonflight manner so that more data would flow out of the rover. This was especially necessary since running software tests on the actual rover presented the risk of having a line of code cause some instrument or item to be turned on, without anyone noticing, which could cause overheating and equipment damage—something that could happen quickly since the rover is so well insulated against the Martian cold. Everybody involved in testing in the High Bay needed to be alert for this at all times. One slip could require fixes with a price tag running to many millions of dollars and, worse, a cost in time lost we could not afford.

The testing revealed numerous software glitches that kept the programmers working late, but the hardware held up. I was beginning to gain a sense of confidence. I think everybody around me was, as well, but nobody spoke of it out loud—at least, not within my hearing. Had we all become too spooked or too superstitious to express our confidence to each other?

By March, it was time to begin packing up all the flight hardware, preparing for the trip to the Cape.

The early morning hours of Wednesday, May 12, 2011. A large truck rolled up to High Bay 1. We watched as the wooden crates in which the descent stage and cruise stage have been packed are loaded on. In the past, flight hardware traveled to the Cape by truck, coast to coast on Interstate 10. This time the ground journey would be considerably shorter, with the truck going only to March Air Force Base (or March Air Reserve Base, as it is now called), some sixty-five miles from JPL. There the crates would be transferred to an Air Force C-17 Globemaster cargo jet. The flight would make a stop in Denver, where it would pick up the heat shield that has been built by Lockheed Martin Astronautics.

Several of us turned up at 2:00 a.m. to watch the truck depart. Before the ramp was closed, I walked into the cargo area, where heavy straps held down the crates. I leaned on one of the crates, and it swayed a little. Disconcerted, I talked to the driver, who said he had driven many items of cargo and never lost anything yet. Somehow that answer wasn't entirely reassuring. But it was now out of our hands.

At just past 1:00 a.m. on June 22, it was the rover's turn. Again the trip would be to March Air Reserve Base and by C-17 to Cape Canaveral. In a tradition I had started three missions ago, I arrived with my trumpet and, as the truck pulled out of JPL south gate, I played "When the Saints Go Marching In" for the handful of people who had shown up. The mood was joyous, and my noisy musical addition was clearly appreciated.

Suddenly it was quiet, and the rover was gone. But for most of us the job was nowhere near over.

While our spacecraft and rover were being delivered, unpacked, and prepared for final testing, the crew at JPL still had its own myriad tests to do, using the test bed rover, plus the vast set of electronic components spread out in the Mission Systems Test Bed. This period turned out to be a time when we could not relax.

Despite all our precautions, one of our routine tests with the drill triggered a major problem. On the test bed rover, the mechanism that sets up the rapid hammering or "percuss" motion of the drill had failed. That failure created an electrical short circuit. If we had not quickly detected the problem, the rover's electronics would have been damaged. Not only would we end up losing the drill percussion capability, but we would also lose the computers and the whole rover.

Could this happen on Mars? We discovered that if the same problem happened again at any time, it would indeed take down the rover. The cause turned out to be something that revealed what could be serious problem with the test bed's drill: one of the bushings—a cylindrical guide for the percussion hammer—was not attached sufficiently, and had fallen off causing an electrical short that affected the whole power system in the rover.

In a tribute to the smarts and dedication of the team, we managed to cobble together a solution in just forty-eight almost nonstop hours. Although we couldn't prevent or fix a broken bushing on Mars (though we would later came up with a way to minimize the chances of it happening), we found that the ensuing electrical damage to the computer and the rest of the electronics could be avoided using a spare electrical switch to route a new grounding wire to the power bus. Whenever we used the drill on Mars, we could command the switch to ground our power bus, and the circuits would be protected if ever the same bushing failure happened on Mars.

We disassembled the test bed rover and tried out this trick. It worked. Great, but by now the flight rover was being readied for launch at the Cape. This fix would require taking the "belly pan" off of the flight rover and performing electrical surgery.

I was faced with a difficult choice. This was September—just two months before our November launch date. Very nearly too late in the game to be pulling off the belly pan cover. If a broken bushing was unlikely, we could risk living with the dire consequences of a drill short. But if not and we did nothing, we could risk losing not only the drill percussion feature but also the entire mission. So, at nearly the last possible moment, we advised project management and ATLO manager Dave Gruel that we needed to open up the bottom of the rover belly to access the circuits. They understood it was necessary and agreed. The ATLO crew at Cape Canaveral, working as fast as they safely could, removed the rover belly pan and added a wire to provide the needed grounding.

It was a bold and rather risky operation. I was jumpy until they were able to email pictures back that showed the ground wire installed, assuring us they had carried out the steps to test it, had reassembled the rover, and were back up to speed.

The drill starred in another major problem. Our SA/SPaH team had provided three drill bits. Two of these were spares: In case the initial bit broke or became clogged or dull, the robotic arm could be sent through a series of steps to replace the bad bit with one of the spares. All three of the bits had been sterilized, then covered and kept clean until it came time to prepare for launch, when the cover was to be removed. That was part of the procedures in place to prevent carrying any terrestrial microbes to Mars. It was especially essential to prevent Earth organisms from contacting water or ice and inadvertently causing them to thrive on Mars. We didn't want to imagine triggering happy shouts from scientists that they had discovered signs of organic chemicals, when in fact the organics they detected had come from Florida. (Anything that touches the Mars soil has to be sterilized. For Curiosity, that was a short list: drill bits and the six wheels.)

During testing of samples taken using test drill bits back in Pasadena, the team had discovered what appeared to be extremely minute amounts of oil coming off the bits, even though the bits had been treated with the exact same care and sterility as the bits on the rover.

Even this tiny amount of organic residue could corrupt the results from the SAM instrument. We wondered if the bits on the real rover had similar contaminates. A few months from launch, at the SAM team's request, our contamination control engineers had gone into the clean room in Florida to take samples to test for contaminates. This involved temporarily removing the bits. They passed the test, but since this was done under extremely clean and sterile conditions, they were not then resterilized with dry heat, as they had been before initial installation.

Under NASA procedures, changes in how the drill bits were handled required permission of NASA's Planetary Protection officer, Catharine Conley. She is responsible for guarding against contamination from Earth being carried to other celestial bodies, as well as for ensuring that potential biohazards are not inadvertently brought back from places like Mars.

The folks at JPL who knew about the rules were not at the Cape but in Pasadena, and they were not aware that the contamination test required removal of the bits, while at the Cape, the contamination engineers did not know that removing the bits—even if they were maintained sterile—still needed approval from NASA headquarters. When we reported the contamination results a few weeks later, Conley was concerned that since we did not follow procedures to resterilize the bits, the drill bits might now harbor terrestrial organisms. She made it clear that no drilling would be allowed in any rock that might contain water or ice. This was not as threatening to the mission as it sounded. It would have been a big problem at other landing sites, but we already knew that Gale Cater was close enough to the dry Martian equator that water or ice, if present, would be many meters below the surface. The only water around was chemically bound with the constituents of some rocks, and was not water that any life could thrive on.

We asked to have the mission reclassified by the Planetary Protection office. There is a category of mission that limits operating to areas of Mars that have "no liquid water near the surface where an Earth cell could survive." We were granted that reclassification, and although the whole matter was embarrassing, the problem was now resolved, and we were good to go.

In the weeks leading up to launch, tensions built as each team at JPL and the Cape ran final tests. EDL and cruise testing on the Mission Systems Test Bed was going hot and heavy. Although the rover teams had access to the rover test bed, that one was far too oversubscribed. The test beds were booked 24/7. The rover team desperately needed another test venue. Fighting for test access once again became intense. Meanwhile we also began around-the-clock team training for all of the people at JPL who would be manning consoles or in other ways playing a role during flight operations. Much of this training needed the test beds, too. Our only solution was to try to improve the way we managed the test beds so that when one team came in for their tests, someone was there to help them set up and get the work done as fast as possible, so the next shift could come in without having to wait.

Finally, more than a decade after conversations had begun about doing serious roving sample science on Mars, the day had arrived. At 10:02 a.m. on November 26, 2011, under clear, nearly cloudless skies, the Mars Science Lab launched aboard an Atlas V rocket from Cape Canaveral to begin its eight-month, 354-million-mile journey.

I observed from the Cruise & EDL control room, where the team was ready to "catch" MSL as it entered space, about an hour after launch. If there were ever a time for something to go wrong with MSL, it would be in the first hours and days after launch. My role for that day was as anomaly lead, standing by to put a team together and come up with a solution in case something went wrong just after launch. Nothing did. We watched images from a camera on the rocket's upper stage as our spacecraft was gently spun up and pushed away on its dark and distant ride to Mars. It was a beautiful moment.

Once again, these next months were not a time we could relax or move on to other projects. The cruise team was learning how to fly MSL from Earth to Mars, practicing every routine procedure and every emergency procedure. Even more challenging, most of the testing of the vastly complex rover had to be conducted while the spacecraft was en route. What's more, the engineers testing the software to be used for landing had a whole list of problems with the design that they hadn't yet solved, and the software had not yet been fully cleaned up to get out all of the bugs. The cruise, EDL, and surface engineering teams would work with the software team in the test beds around the clock to find and fix problems. It would be months before they could begin beaming through space the final version of the EDL software and, after landing, the software that Curiosity would need on Mars, to be uploaded on the rover's computers.

In a perfect world, I would be relaxed and confident as MSL soared through space on its way to Mars. But I know better. It wasn't just my imagination picturing the worst. It had actually happened. Eight years earlier, the twin MER spacecraft, Spirit and Opportunity, were arcing across the solar system to Mars. As Spirit neared its destination, we suddenly felt that getting them designed, built, and sent on their way in only three years may have been the easy part.

The first sense that Spirit's landing might not be routine had come on Christmas day, about a week before its scheduled arrival. The orbiting Mars Global Surveyor had spied a massive dust devil in the southern half of the planet—far from Spirit's landing site but close enough to give us the willies.

Later that day, we waited for confirmation that Great Britain's first Mars lander, Beagle II, had arrived safely. The confirmation never came. It seemed a bad omen.

The following Tuesday, December 30, four days before Spirit's landing, I set up a meeting with the EDL team to review the final results of the last EDL tests that we had run on the cruise and EDL test bed in our lab. I expected little other than "Check, check, check" from the results. In fact, that's how the meeting went until the review of the very last tests.

Avionics expert Jim Donaldson had apparently asked for an "EDL-run"—a software run-through of the entire EDL program—and requested that it be run on what we call the software test bed instead of the test bed usually used for this type of run, the EDL test bed. Everything worked perfectly with two rather important exceptions: the airbags did not inflate and the rockets did not fire. The test was run a second time, with the same result.

Jason Willis, the EDL systems engineer, sequence designer, and EDL test lead for Spirit and Opportunity, listened to the report and was fuming. This was his area of responsibility, and no one had told him that these tests would be run. He was convinced that the failure was due to problems with the software test bed and had nothing to do with the actual spacecraft, but now, on top of all of his EDL preparations, I was asking him to assemble a team and prove it. He was not a happy camper.

Within twenty-four hours, he had found there was nothing wrong with the software test bed. Within another twenty-four hours, he and the EDL test team were able to reproduce these scary, mission-threatening results. Something was definitely wrong. What we finally discovered was that messages that controlled the precisely timed pyrotechnic explosions that were at the heart of all of the EDL events were being lost—they had been sent from the rover electronics but not all of them were received by the electronics in the lander that would initiate the pyrotechnics.

Jim Donaldson dove into the lander's circuit design with the original circuit designer. It took them only a few hours to discover an electronic design error that had never come to light before. As long as the electronic clock inside the rover electronics ran faster than the clock inside the lander, the message to enable the triggering of these explosives would get through as designed. But if the clocks inside Spirit and Opportunity drifted even slightly the other way around, the message would not get through, and the tiny pyrotechnic explosions that inflated the airbags and fired the rockets would never occur. The rover would crash into Mars. For some reason the clocks in the software test bed drifted in one direction while they drifted in the opposite direction in the EDL test bed.

When we looked at the clock drift on both Spirit and Opportunity, to our dismay we discovered that they were both more like the software test bed than the EDL test bed. In fact, that clock drift difference was what had motivated Jim to ask for the EDL test run on the software test bed in the first place. There was a growing chance that neither of the two spacecraft would survive.

It was now a little over twenty-four hours from landing for Spirit, and I was beside myself. I pleaded with Glenn Reeves, the chief engineer for the rover's software, to see if his team could figure out changes to the software that would work around the hardware bug.

He shook his head. "Rob," he said, "it's too late to fix the software."

"Then we're sunk." My head was spinning.

"Don't panic," he told me. "We'll send the signal to enable the pyros manually. Tomorrow night, a few hours before landing, we can send low-level commands to the rover that will do what the software was supposed to do."

I understood what he was saying, and knew that his plan would work. What a huge relief. We had been so close to losing the entire mission, all because of one small circuit design bug and a clock with a slight time drift.

That night, the night before landing, we successfully tested EDL three times using Glenn's trick. It worked.

* * *

Those memories of Spirit and Opportunity had washed through my mind when Mars Science Lab had launched so smoothly. But it had been en route for only three days when I got an alarming call from the cruise Mission Support Area. The rover's computer had unexpectedly rebooted just minutes after the cruise team commanded it to begin looking for stars.

The rover computer stores a complete star catalog. Its star scanner looks at the sky, locates stars, and from their positions calculates its orientation in space—all of which is quite amazing when you stop to think about it. But now, for some reason, the rover computer crashed when it tried to see stars. Why? We had no idea. A software bug, maybe? This was something we had never seen before.

Now the spacecraft did not know where it was pointing. It knew where the sun was but that was all. If MSL couldn't see the stars, it couldn't determine which way to fire its thruster when the time came to correct its trajectory, and so would never find its way to Mars. The thought of the spacecraft heading off to some remote part of the heavens was sickening. The good news was that we had time. The spacecraft was headed in the right direction and we had a couple of months before we needed to do a trajectory correction maneuver.

Brian Portock's Cruise Ops team sent a command ordering Curiosity to try running a star search again. Instead, its main computer began *rebooting*. What it sent back was "autopsy data"—the computer reporting that it had received an illegal instruction.

I put together a Tiger Team to figure out the problem and come up with a solution. It took weeks, anxiety-inducing weeks, before they were able to confirm that it did not appear to be a software bug. We reviewed the data from the rigorous testing done earlier and consistently found that the software had worked perfectly.

It took almost eight more nerve-wracking weeks before someone finally stumbled on the answer. Dan Gaines, an incredibly smart young member of the software team, recalled a glitch he had noticed early in our testing that had made him at the time wonder if it could possibly be a hardware defect. Following up on his hunch, we were able to find spare computers in the storage area of our test bed and near-duplicates of the flight computer cards. When we loaded the computers with the cards, to our amazement, some of those computers duplicated the problem that had happened to our MSL spacecraft.

We were shocked to find that the fault was not in the software but in the computer itself—in fact, in both the main computer and the backup. The computer processors had a subtle and quite old design defect that prevented them from running certain combinations of instructions. It didn't seem to make sense. None of us had ever encountered a problem like this. Moreover, these computers had been used by many other spacecraft and had decades of flight time with few problems.

Understanding the rough cause was a great relief, but we had no idea whether the problem could be fixed or if we could even figure out what was going on inside the computer chip. Digging deeper, we learned from the vendor that the way we were using the computers was slightly different from the way anyone else had ever used them. After running detailed simulations of our computer circuits, they discovered that the way we were running the rover's computer left us vulnerable to a newly discovered tiny design defect, a small electrical glitch unbelievably lasting only some 300 to 400 trillions of a second, that under these certain rare conditions caused the computer to misexecute an instruction.

This wasn't something we could fix. We discussed the problem with the company that had provided the computer, and its engineers dove in to find a solution. We were anxiously counting down the time remaining. The kind of small course corrections that would be needed could wait until fairly late in MSL's journey. We waited. And waited.

I began to wonder when it would be time to panic. Finally, the vendor came up with what he assured us was the solution. All we needed to do was to change the software to avoid the use of certain memory control functions in the computer chip. If we did that, we wouldn't prevent the glitch, but we would avoid the problems that the glitch was causing.

It took the team days to configure the new software, but it turned out that the manufacturer was right. At last MSL could see the stars again. Finally, the vehicle turned to make the needed course corrections.

In fact, we were lucky. The experts who had programmed the Atlas V 541 rocket that put our spacecraft onto its initial path had done a fantastic job of sending it on its way with great accuracy, so that the amount of correction we now needed was very small. If the path of the launch vehicle had been just a little less accurate, the solution to the glitch would have come too late to make the degree of correction needed, and MSL would have been on a journey to oblivion.

When you build something new and complicated—especially with a mission like this one that had so many problems—you're constantly waiting for the next calamity. Still, this computer experience upped the ante and made everybody nervous. If the spacecraft suddenly could no longer see the stars, what other bugs were lurking in the system that we hadn't found yet, bugs that could sink the whole mission? The star sighting was comparatively so straightforward and EDL so vastly complex. The experience gave all of us a sense of foreboding.

CHAPTER 16
Gremlins

On an earlier mission, I had learned a painful lesson about team psychology. We arrived at the launch pad at the Cape having had little time for rehearsing the launch. During the countdown, I could see that everyone was growing more and more panicky because of all the things we had not had time to discover about the launch preparation and procedures.

On launch day, our team clumsily and nervously marched through the countdown sequence. With only moments to go, Launch Control advised that a fishing boat had entered the restricted launch area. The Range Safety Officer ordered the launch cancelled until the following day.

That scrubbed launch had provided just the practice we needed. We spent the rest of day revising our shaky procedures, followed by a good night's sleep.

When we assembled in the control room the next morning, the chemistry was entirely different. Everyone was much more relaxed and confident. I caught on: We need to learn how to stand back and come to terms with the "situational awareness." What is it doing now? What state is it in? If you rehearse rigorously enough, on the real day any problems that arise are not nearly so wrenching. On all future launches, we would make sure the team was well prepared.

With a few months to go before landing, we were finally getting to the point where the rover's EDL software appeared to be working pretty flawlessly on the test bed. But it's not possible to test it hundreds of times. EDL begins when the rover is still tens of thousands of miles away from the surface of Mars. Testing a single run-through of the EDL software was taking an entire ten-hour test bed shift. Testing the entire process for landing took much longer.

The whole process from cruise to surface operation requires the coordination of at least five separate teams. There's a lot for these team members to do in the last few days leading up to landing night. Here's just a sampling:

- Lou D'Amario's navigators must figure out where the rover is relative to Mars and estimate precisely where it is going to enter the Martian atmosphere on landing day.
- Brian Portock's cruise operations team (Cruise Ops) has to ensure that the rover, descent stage, and cruise stage are all warmed up, checked out, and ready for landing, and that the final course corrections are performed flawlessly.
- Adam Steltzner's EDL operations (EDL Ops) team must make sure the hundreds of parameters that the EDL software will need for flying the lander precisely into Gale Crater (things like time of arrival, the best estimate of the rover's position and speed relative Gale) have been checked and sent.
- Brian Schratz and Peter Illsley's EDL communication team needs to be ready to coordinate EDL displays, the Deep Space Network (which transmits our radio signals to and from the spacecraft) and three orbiters (Mars Odyssey, Europe's Mars Express and the Mars Reconnaissance Orbiter) to ensure that data will stream back to JPL during EDL.
- Jessica Samuels, Jennifer Trosper, and Rick Welch's rover operations team must be ready to assess the health of the rover after landing.
- Mission planners, sequence integrators, telemetry analysts, and ground system engineers must be ready to prepare the commands to be sent and process the data that continually flows from the rover.
- Finally, the public communication team needs to be set up to route live TV coverage to the outside world.

This highly complicated, carefully choreographed operation, like all big operations, needs to be rehearsed. We had scheduled a series of what we called Operations Readiness Tests (ORTs) using real orbiter radios, simulated Mars orbiters and a simulation for the Deep Space Network. The Mission Systems Test Bed is an ideal stand-in for our real spacecraft, with complete sets of electronics for the rover, cruise, and descent stages as well as equipment the software uses like thruster valves, and star and sun sensors.

We could even simulate the thirteen-minutes-plus it takes for a radio signal moving at the speed of light to travel from Earth to Mars. We could also simulate that same delay for a message to return from our spacecraft near Mars to Earth. That way when the team in the Cruise control room sent a command, it would look to them as if it took thirteen minutes to get there and another thirteen minutes for the response to come back.

One of the biggest challenges is figuring out how to trick the test bed "spacecraft" into responding as if it really were in outer space. In many cases we simply had to tell the Ops team to ignore the fact that the spacecraft reported sensing Earth's gravity, or that the cruise propulsion fuel lines were at room temperature. But in many other ways it can be hard to tell that the test bed is not a real spacecraft headed to Mars.

Our first two Operational Ready Test rehearsals involved almost everybody currently working on the MSL project. These tests would be what we like to call "nominal" ORTs—(again, meaning "nothing wrong, everything as it should be"). These rehearsals went fairly well, though there was a lot of rough around the edges.

Nominal ORTs are boring. Unless someone accidently pulls the plug on the test bed, there's not a lot of drama, nor is there a big need for the team members to think independently and make decisions. The engineers at the consoles are particularly bored. They had spent the last few years racing to get the spacecraft designed, built, and tested. This Ops stuff was dull and mechanical—until something goes really wrong, and then it flips into becoming *too* intense.

That's the reason for the real meat of these drills, which we call "off-nominal ORTs." These are tests where the team is presented with simulated problems—things going wrong with the spacecraft.

The people who conceive the "things going wrong" to present to the team are known at JPL as "gremlins." For the final off-nominal ORT, I was to be chief gremlin. The tests would cover four days prior to landing and the first

four full Martian days of operations and would call for actions involving the
cruise team, the EDL team, and the surface team.

There is something quite satisfying about being in a position to intentionally
trip up your friends and colleagues without consequences. Once the control
room teams had settled into the routines and seemed as comfortable as the
pilots of an airliner, I began to turn up the heat.

They say that the most important part about life is knowing when to show
up. At the outset, everyone needs to know when to come in to the Cruise
control room and where to sit. There is also a separate control room in a
nearby building for the rover operations. Each person is handed a three-ring
binder full of procedures to guide him or her in every step. The flight director
keeps the tempo and makes sure that everyone keeps up and takes the correct actions.

No one wants to take a chance that the team might get the real spacecraft
confused with the test bed and accidentally send to the real rover a command
to take over control and start to execute entry, descent, and landing. To avoid
any possibility of this, the displays used for the rehearsal drills have been
designed to look markedly different from the displays sending commands to
the real spacecraft.

During this final, dreaded ORT, we would be watching closely to see who
on the team would keep a calm head when others panic. Who are the best
problem solvers? Who are the people we will want to turn to in an actual crisis? My gremlin crew of about half a dozen engineers and I had spent some
three weeks cooking up challenges for this final ORT.

Since bad things seem to happen in bunches, we were not too concerned
about having too many things go wrong, at the same time making sure that
some of the problems were evenly spread out among the teams. We needed
to make sure no team was deluged but that everyone was kept busy and that
there was plenty of need for interteam communication.

On the first day of this ORT, the team arrived for what they were told was the
morning of August 2, only a few days before landing. They settled into the routines. I thought it would be good to start the morning with a red herring. I
asked JPL's space radiation expert, Martin Ratliff, to send an urgent email to

the cruise mission managers with news of an impending huge "coronal mass ejection" from the sun: a huge solar flare heading toward Mars.

Martin's warning to the team was a heads-up. A genuine large solar flare would bring a real risk that the rover electronics would suffer "bit flips," with ones and zeros stored in the computer getting scrambled. (In fact, this actually did happen to both Spirit and Opportunity a couple of months before they landed.) The team needed to be ready, prepared to deal with possible rover computer resets or other unusual actions.

They seemed to be handling everything fine until about 2:00 p.m., when the navigation team met to discuss the last chunk of radio navigation data that had arrived from the (simulated) Deep Space Network. The team had noted earlier in the day, before the solar flare, that the spacecraft seemed to have started to decelerate slightly. Something extraordinarily lightweight was pushing on it, with a force no greater than the weight of a single piece of tissue paper in the palm of one's hand.

I thought, "Damn, those navigators are good!" I had expected it would take them at least until late afternoon to notice. Still, as I sat in on their afternoon status meeting, they really didn't think a lot of it. It wasn't so large as to make a difference. Besides, as the navigation team chief, Lou D'Amario argued, if it was a constant force, their software would take it into account when the time came for the last trajectory course correction in a couple of days.

What they were dismissing as unimportant was in fact significant. We gremlins had created a daunting scenario. We set up inputs to the test bed and to the navigation team that simulated the effects of an extremely tiny micro-meteor impacting the bottom of the cruise stage, punching a nearly microscopic hole in a fuel line. The flow of fuel from this leak, small as it was, would be enough to push the spacecraft off course.

To up the ante, we simulated a situation in which the leaking fuel would freeze over the tiny hole for half a day, then melt; the leak would start again, only this time in a slightly larger stream, and one that would affect the 2 rpm spin rate of the whole three-ton spacecraft. There would be no way that the navigation team would be able to make reliable predictions on the spacecraft's actions over the following days.

To make sure the cruise team noticed, we adjusted the fuel pressure every few hours so that after a day or so the leak would become obvious to them.

My fellow gremlins and I hoped that this minute leak would end up pushing the spacecraft off course just enough to force the team to choose between two possibilities, both with potentially grave downsides: Let the spacecraft attempt to "fly out" this huge targeting error, making its own large course corrections during the guided entry phase of EDL and risk causing the vehicle to land far outside Gale Crater. Or risk a possible explosive loss of propellant by using the leaky cruise stage propulsion system to make the needed course correction. I had hopes that this conundrum would put pressure on the navigation team, the EDL team, and the cruise team, forcing tough project-management risk decision choices. It did.

Late that night, we threw another problem at the teams, making it appear that a solar particle had created a memory error in the rover's "prime" (primary, in this case, the A-side) computer. The teams saw that the prime computer had gone through a reset, causing the backup computer to take over. But the backup computer didn't yet have all of the files needed for a safe landing. The team was faced with a choice: Swap back to the prime computer and assume that it would resume working properly, or send the final files for EDL to the secondary computer and allow it to run the landing. Since they couldn't be certain if the memory corruption of the A-side computer was permanent, they chose to load the files to the B-side. Good call.

The next morning, we provided data to indicate that the temperature sensor on the rover's high-gain antenna had failed; this is the antenna used by the rover to listen to the day's instructions from Earth after it wakes in the morning. Without this sensor, the antenna's temperature control software would not work and the antenna would be far more difficult to point. I was happy to see that the Rover Surface Ops team was on it and was instantly worried that they would not be able to get commands sent to the rover via the high gain antenna. The rover would become deaf until somebody or something swapped to the backup sensor. By noon they all agreed that the surface team would have to deal with it after landing. Another good call.

Finally, in our simulation, it was the night before landing day. By then most of the teams were pretty tired. So was I.

The cruise team had finally diagnosed the fuel leak, the first problem we had thrown at them nearly four days earlier. They even correctly figured out that the leak we invented was coming from the under side of the cruise stage near the fuel lines. I thought that was pretty smart of them.

By now the simulated rover was about 6 miles (10 km) off course. I assumed that the team would conclude that it was probably safe to use the cruise propulsion despite the leak. To my surprise, the debate became heated.

I watched quietly as they discussed the down sides of using the leaky equipment. I was amazed at all of the negatives they recognized—in particular, that there was a possibility using the cruise thrusters could cause the small leak to pop wide open, sending the spacecraft spinning wildly out of control.

Strongly in the "no thrusters" camp was Dara Sabahi. He forcefully argued that the choice should be to rely on the EDL capabilities and not tempt fate with the leak. It might be that the EDL guidance would be able to fly properly and stay on course to the landing site in Gale. We just didn't know. On the other side were Adam Steltzner and his EDL team, who were beginning to have doubts about their system's abilities to handle such a large discrepancy between where MSL would enter the atmosphere and where we wanted it to land. They had never counted on being asked to deal with such a large position error and weren't certain if the entry guidance design could handle it.

Dara vented that they *should* know what would happen. Adam agreed, acknowledging that this was part of their effort that his team hadn't finished yet.

Making matters worse, my gremlins and I threw in a mean little trick. We announced to the EDL team that one of their members, Dan Burkhart, suddenly had to go home on an urgent personal matter. Dan was the main guy responsible for translating navigation data into commands that would tell the rover precisely where we thought it would be just before it arrived, how fast it was going, and precisely where we wanted it to land inside Gale Crater. I was sure that Dan's boss, Adam, would have a backup person who could step in.

Adam came racing in to find me.

"Dammit, Rob, what are you *doing*? We might as well just quit now. Dan was just about to deliver the file! It's not realistic he would leave now in real life!"

"Yeah, well, he could get sick," I countered. "Don't you have a backup?"

"Yeah, but he finished his shift and has already gone home. He lives a long drive away, and he isn't answering his phone. This whole ORT will be a disaster unless you let Dan do his job," Adam insisted.

I asked, "If you were in this situation in real life, would we lose the mission?"

"Maybe. Maybe not," Adam answered.

I agreed that if he couldn't find anyone who could build the command files, I'd let him have Dan back so the ORT could be completed. But it wasn't necessary. With barely a moment to spare, Adam and his team figured it out over the phone with Dan's backup and got the files built and sent to the test bed spacecraft without Dan.

The next morning, the teams were still split on what to do about the leaky fuel. Project manager Pete Theisinger and his deputy, Richard Cook, joined the deliberations, eagerly treating this as if it were real life instead of a simulation. Dara made a rational, compelling recommendation that they not take chances with the thrusters. But the last thing Pete wanted was to have a failure during EDL caused by us knowingly flying off course, running the risk of hitting the rim of Gale Crater. It felt too much like the Mars Climate Orbiter loss. After going around the room for more opinions, Pete reluctantly made the decision to use the thrusters for a course correction.

Finally our simulated EDL night arrived. It would be a long night. To my surprise, JPL's lab director, Charles Elachi, and the president of Caltech, Jean Lou Chameau, both showed up and took their seats in the back of the control room, just as they would during the real EDL night. Their presence would add to the pressure and the drama.

Even the JPL camera crews and the NASA TV interviewers were on hand, carrying out their roles just as they would during real EDL. After Pete and Adam had given their interviews, I took my seat in front of the camera and pretended to be excited about a successful landing "tonight." Al Chen, the EDL systems engineer, was announcing MSL's progress as each event took place. I could tell that the atmosphere in the control room was tense, even though the team was cheering as the EDL actions were clicked off one by one, just as if this were real instead of a simulation.

I jumped into my Honda and quickly drove a simulated 150 million miles over to the test bed, donned a white lab coat, and joined my fellow gremlins, excitedly waiting for the rover to begin its automatic sequence of events. I arrived just as the test bed vehicle was entering the Martian atmosphere.

Before the test started, I had asked the gremlin team to install a toggle switch on one of the circuits that attached the rover's computer to its UHF radio. As soon as the rover had separated from the descent stage and the rover wheels had been deployed during the sky crane maneuver, I flipped the switch.

In my scenario, the pyrotechnic shock from the rover wheels being released had opened a single pin in the electrical connector between the rover's radio and the rover's computer, causing nothing but zeros to be transmitted. Nothing that couldn't be fixed, but the uncertainty it would create in the control room . . .

In the control room, just as the rover was about to land—nothing. It all stopped. The room went silent as they waited for Al to continue. But Al was saying nothing. Because of my flipped switch, the digital stream from the (test bed) rover to the (real) MSL control room had stopped. The room went sickly silent when the data link disappeared.

I yelled my congratulations to the gremlins, jumped into my car, and raced another 150 million miles back to Earth. By then an anomaly team had been assembled in the conference room adjacent to the control room. They needed to get to work to see what might be gleaned from the data that they did get before the data stream was lost. The rover was scheduled to go to sleep just a few minutes after landing and wake up again in a little over an hour, when the next Odyssey orbiter would come into position to relay Curiosity messages to Earth.

At this point, no one knew if our simulated rover had landed safely or not. Did loss of data mean that the signal had dropped out, or that the rover had crashed? What could they know for certain about what had happened? It suddenly became clear that we really did not have a plan for how we would conduct an investigation or what or when the public would be told about our having "a bad day." Theisinger and Cook spent about thirty minutes with Elachi and Chameau and our public affairs folks discussing what we should do in this situation, as I quietly smirked in the back of the room. If ever there was a good time to dream up a bad day contingency plan, this ORT would be as good as any. But I shouldn't have been smirking: I would learn a few days later that I was being tasked with creating that plan.

What to do next became a big question. In principle, the surface team might have made a lucky guess that the UHF radio had been broken on landing. Since the rover had a backup radio, they could send a command to swap radios, transmitting over the X-band link that we use for routine telecommanding. This would have instantly fixed the problem the gremlins had created.

But there were too many possibilities. What if the rover's radio was fine but the computer failed and did not swap to its backup? What if the rover landed in a crater that blocked the radio? Although they knew there were subtle hints in the radio data collected during landing that something was up with the radio, perhaps my gremlin team had made it too difficult for them to guess which would be the right thing to do.

It was hours before they finally did send the command to swap radios. But the command didn't work, which made it look as if perhaps the rover hadn't survived the landing after all. That caught us gremlins off guard. Switching to the redundant radio was supposed to solve the problem.

I was starting to get nasty looks. One of the MSL crew even texted me a short video showing me sitting at a conference table as a bomb came bouncing into view and exploding in my face. Ouch! (The wonders of some of those two-dollar apps.)

After yet another quick trip to "Mars" (the test bed), I discovered we had forgotten to plug in the cable to connect the backup UHF radio on the rover. That was embarrassing.

About two hours later, while the rover was automatically taking a nap, my gremlins snuck in and wired it back up properly. At the next communication window with the orbiter, it worked. The team yelled when they finally got confirmation that the rover had survived landing after all. Nearly twelve hours after landing, the surface ops team could finally begin to get data and pictures from Mars. They spent the next day trying to configure the rover to be useful after the "failures" that my fellow gremlins had inserted during the final three days of cruise.

At the end of the ORT a few days later, they were all tired, and I was exhausted. If I had had a Facebook account, I would have expected a few hundred new "enemies" on it.

It was interesting to me, though, that despite everyone knowing this was only a simulation, it all felt so real that emotions were running high and people were genuinely frustrated, depressed at near failure, and stunned at how hard it was to operate this beast.

Later, a journalist who had taken an interest in these Ops Readiness tests, Amina Khan, wrote of me in the *Los Angeles Times*, "he seems to take an unholy glee in his stint on the dark side." Yes, yet in the end I was proud of how well the whole gang had worked together and solved all the problems. My fellow gremlins and I had thrown a mind-numbing array of challenges at the team and they passed with flying colors. The lessons we all learned over that intense week in May did more for the team than anything they could have learned from studying the documentation on handling failures and emergencies. It also psychologically steeled them for the struggle that we all needed to be prepared for if this mission were to be a success. In particular, it would vastly reduce the anxiety during the actual descent and landing.

After this week, could the real landing day be any worse? I hoped not.

CHAPTER 17
On Mars!

Just turned off the uplink signal—the spacecraft will operate autonomously through landing. Take care Curiosity. Godspeed. #msl #jpl #nasa

That tweet was sent by JPL systems engineer Bobak Ferdowsi. From this point until Curiosity was on the surface of Mars, there would be no need for commands. For better or worse, the computer deep inside MSL's rover had been given complete control of Curiosity's destiny. We were onlookers, with no more command over the craft than the worldwide audience of television viewers.

Bobak had made it a practice of changing his Mohawk hairstyle for each big event on each of the spacecraft projects he had worked on. This time he'd given his teammates the opportunity to vote on how his Mohawk would look, and they had decided on white stars and red and blue spikes. On landing night, because he was seated close to the television cameras sending images around the world from the JPL control room, his distinctly unusual hairstyle made him a focus of attention. Even President Obama, in his praise after the landing, mentioned the popularity of "Mohawk Guy." His mane instantly made him both famous and our team mascot. If anyone on MSL could represent the team with pride, Bobak could, and he did. He was even invited by Mrs. Obama to be her guest at the president's State of the Union address before Congress.

Excitement over the approaching arrival on Mars had been building. NASA's Twitter account providing details on Curiosity had attracted more than 100,000 followers, while *Seven Minutes of Terror*, a nerve-wracking video posted on YouTube by JPL's John Beck, had already been viewed more than half a million times. We all hoped that those fans around the globe were not going to be disappointed.

The intense EDL phase of the mission begins when the spacecraft reaches the top of the Martian atmosphere, traveling more than 13,000 miles per hour (5,900 mps). It ends after those seven minutes of terror, with the rover stationary on the surface.

The previous Monday evening, six days before landing, MSL was still more than 1.2 million miles (2 million km) from Mars. EDL systems engineer Martin Greco prepared and sent out a simple digital message to start the rover's EDL software countdown to landing. From then on out, Curiosity would do its best to land on Mars without our help. The message read simply:

DO_EDL

With this command, the rover was told to take control.

On landing night, August 5, 2012, the heart of the complex EDL sequence began. Ten minutes before the spacecraft arrived at the top of the Martian atmosphere, the cruise stage separated from the aeroshell. A minute later, thrusters fired, turning the aeroshell so that the heat shield was facing Mars. Ballast aboard the spacecraft was released, offsetting the center of gravity so the heat shield tilted down against the oncoming airstream; this was the winglike attitude, providing lift that the entry guidance software would use to fly toward Gale Crater. The craft was now on a shallow 15.5-degree descent trajectory, flying like an airplane, with the computer serving in its fleeting career as an automated hypersonic test pilot.

Reaching the entry point at the top of the atmosphere, the spacecraft was about 78 miles (125 km) above the surface. Over the next two minutes, the computer in the rover tracked the gradual decrease of speed as it slowed from the drag of the heat shield.

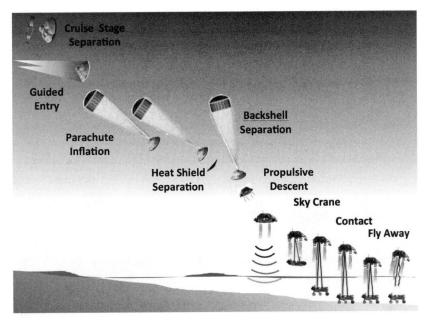

FIGURE 17. MSL's entry, descent and landing (EDL) sequence of events, starting with cruise stage separation about ten minutes before entry into the Mars atmosphere and ending at landing about seven minutes after entry. An animation of this sequence is online at CuriosityRover.info. (Courtesy of NASA and JPL/Caltech).

Using the same strategy as a glider pilot approaching the runway, extra speed was being maintained for safety: It would provide freedom to maneuver and make corrections in case of an unexpectedly low or high air density, high-altitude winds, or a similar surprise. The unneeded speed was shed quickly, as the computer sent commands for rolling the aeroshell to the left or right, banking as needed. Eight large thrusters controlled the aeroshell's roll orientation, giving her the ability to fly large, lazy S-turns in the sky as she slowed with the drag of the heat shield and her computer continued to adjust the flight path. There was no danger of landing on top of Mount Sharp or on the mountainous rim of Gale Crater.

The Martian atmosphere has enormous variability compared with Earth's. The air pressure at a given location is known to swing as much as 10 percent in a single sol. That would be like the air in Santa Monica becoming as thin as Denver's in a single afternoon. We could not know in advance how thin or thick the Martian CO_2-laden atmosphere would be. Cold air would be settled low and dense on the surface; if warm, it would be high and thin. Without the lift from the heat shield and the slow bank maneuvers to guide it to the landing spot, the uncertainty in the air density alone would result in a huge range of more than 75 miles (120 km) in the area where the rover might ultimately land.

Within two minutes, the heat shield glowed at temperatures of thousands of degrees while the rover deceleration force grew to an immense 15 Earth gees of force. At that deceleration rate, a 150-pound (68 kg) astronaut would feel as if he or she weighed 2,250 pounds (1,020 kg) and, without a special anti-gee suit, would likely pass out.

The craft was now flying horizontally about 6.2 miles (10 km) above the surface and nearing the starting edge of its landing ellipse. The speed had slowed to 1,000 mph (470 mps)—still fast, nearly twice the speed of sound, but now slow enough for deploying the huge supersonic parachute that had given us so much anxiety.

FIGURE 18. Rocket thrusters orient the entry vehicle as it approaches the top of the Martian atmosphere. They continue to fire to ensure that the vehicle stays on track toward the north plains of Gale Crater as it decelerates. (Courtesy of NASA and JPL/Caltech)

FIGURE 19. At just the right moment, the supersonic parachute is launched from a mortar cannon at the top of the vehicle, slowing the vehicle to less than 200 mph (322 kph). (Courtesy of NASA and JPL/Caltech)

Next, the ballast weights attached to the backshell of the entry vehicle were launched overboard in a maneuver we call "Straighten up and fly right." No longer canted to produce lift, the rover now fired the pyrotechnic device that lighted off the mortar cannon containing the 100-pound chute. Within two seconds, a disk of orange and white nylon fabric, 70 feet (21.5 m) in diameter, was violently dancing behind. It quickly slowed the rover.

With the parachute taking over the job of slowing the descent, the heat shield was allowed to separate, and it fell away.

Almost everything to this point had followed the tried-and-true landing architectures used on all six successful US Mars lander missions, starting with the venerable twin Viking landers in 1976.

At 200 mph (320 kph) and 5 miles (8 km) above the surface, the radar mounted on the descent stage began to sense the speed and distance to the ground. When about 0.6 mile (1 km) above the surface, the rover with its

FIGURE 20. About a mile (1.6 km) above the ground, the rover and its "jet pack" are released, and the descent engines fire to continue the slowing. A multibeam radar on the descent stage informs the rover of its height and speed. (Courtesy of NASA and JPL/Caltech)

descent stage backpacked on top fell away from under the backshell and parachute, which continued to descend to the surface. The eight large throttled rocket engines fired and quickly made a divert maneuver to avoid hitting the backshell and the parachute, and then slowed the craft until all horizontal motion across the surface stopped. Curiosity was now headed straight down at about 70 mph (113 kph) and was continuing to slow.

From here on, what made MSL's landing system so obviously different was the sky crane maneuver that was now about to begin.

By this time the craft had slowed to less than 2.5 feet per second (0.75 m/s). Only 75 feet (23 m) above the ground, the rover released herself from the descent stage onto her three nylon and Vectran bridle ropes, which paid out gradually along with an electrical cable that carried the communication between the rover's computer and the descent stage electronics.

As the rover lowered, the small explosive charges released the rover's six wheels from their stowed position, and they snapped into place. The rover now hung 25 feet (7.5 m) below the descent stage as it continued its slow approach.

Soon all six wheels made contact with surface. As the descent stage continued to drop, the rover sensed the change in the rocket thrust and recognized that the rover had landed.

FIGURE 21. Seventy feet (21 m) above the surface, the rover is released from the descent stage at the start of the sky crane maneuver. (Courtesy of NASA and JPL/Caltech)

FIGURE 22. The rover gently lowers to the surface. Once the rover's weight has been removed from the descent stage's load, the rover instructs the descent stage to fly away. (Courtesy of NASA and JPL/Caltech)

FIGURE 23. The MSL as it descends on its parachute into Gale Crater on August 5, 2012. The Mars Reconnaissance Orbiter spacecraft took this image as it orbited overhead. (Courtesy of NASA, JPL/Caltech, and the University of Arizona)

In the final step, the rover told the descent stage to stop, the connecting cords were severed, a final command to "Fly Away" was sent to the descent stage, and the last electrical cable was cut. From there the descent stage turned away from the rover and throttled up its rocket engines, sending it on a short voyage to crash to the surface at a safe distance from the rover.

Curiosity had landed. She transmitted a report that she had touched down, acknowledging that she was safe upon the Martian surface.

Before and after the successful landing, reporters, friends, and JPL associates asked me what I was most worried about as landing day approached. I told them I was perfectly comfortable with the novel, radical EDL. In my opinion, it was the best Mars landing system yet. My big worry was about a single operation, the one that begins with touchdown and a command triggering the rapid but extremely intricate series of events that cuts the rover free from the descent stage and commands the descent stage to do its flyaway so it doesn't come crashing to the ground on top of the rover.

The operation involves many commands and operations that need to work fast and in exactly the correct order. It was the product of dozens of compromises that we could never test completely from start to end. Yet it all worked.

Within minutes after the landing, images taken from the front and rear Hazcams arrived at JPL, giving startling visceral proof to our anxious team that Curiosity had indeed landed on Mars. Our first views clearly showed Mount Sharp looming to the south, awaiting our visit. My relief at seeing these images taken seconds after landing, definitively telling us that Curiosity had settled to the surface intact, will forever be one of the high points of my life.

The teams of scientists who have instruments on Curiosity began their efforts years ago. Now they were tantalizingly close to being able to begin studying the nature of Mars and searching for conditions that might have

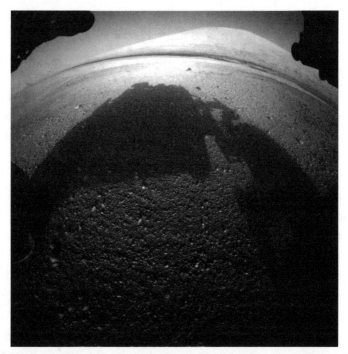

FIGURE 24. This is the first image captured by Curiosity, taken minutes after touchdown by the fish-eye lens of the Hazcam, located near ground level, showing Mount Sharp in the distance. (Courtesy of NASA, JPL/Caltech, and Malin Space Science Systems)

been conducive to ancient life. But not yet. This had to be immensely frustrating. It had taken ten years and some $2 billion to get here. The rover wouldn't
begin roving until it had been thoroughly prepared and tested in about every
conceivable way.

In its state after landing, the rover wasn't yet in condition to do much. The
computers had only enough memory to store the software code for the journey,
the landing, and some rudimentary rover operations. A few sols after landing,
new software for controlling Curiosity and all of its instruments started to be
uploaded, a package of files containing *3.5 million* lines of new code that took
about four days to transmit. The software for all the operations in flight and for
entry, descent, and landing were then overwritten with the new code, endow-
ing Curiosity with the skills that would be essential for exploring Mars.

Once the new code was loaded into both the primary computer and the
backup, the next order of business was to complete an extensive health check
of the scientific payloads and the robotic components of the rover. The surface ops team found that all but one of the instruments had survived intact.
We discovered soon after landing that the REMS instrument, the Spanish
weather station mounted on the rover mast, had suffered some damage. Data
from REMS confirmed that one of its two wind sensors had failed. Pictures
taken from the engineering cameras on the top of the rover's mast showed
pebbles kicked up from the descent engine's blast during Curiosity's landing
strewn across the top deck. Apparently, a pebble had slammed into one of the
two tiny and delicate wind sensors.

I took this as a fault in communicating: We assumed that all the scientific
teams had been advised that the instruments might be pelted during landing.
We had missed an important item, and it was our fault for not making certain
the Spanish team knew there was a danger. However, their second wind sensor was undamaged, so would still be able to gather some wind data. Even so,
I sought out the Spanish team members who were at JPL and extended our
apologies, expressing my chagrin.

Other than the damage to the wind sensor, every other instrument passed
muster at 100 percent, operating as well or in some cases even better after
landing on Mars than at any time during the years of testing on Earth.

* * *

Up until this point, most of the scientists involved with Curiosity had been focused on their own instrument and experiments. But soon their attention turned to watching the rover slowly take its first baby steps in our grand plan to explore Gale Crater: They all waited eagerly to hear a report on Curiosity's first test drive on Mars.

It took place about two and a half weeks after landing, on August 23. For all the anticipation this event had generated, the first drive was a very short trip. The mast, capable of turning 360 degrees, was commanded to turn one-third of the way, and did without problem. Then Curiosity carried out instructions to move forward about 15 feet (4.5 m), and then back up about half the distance. She trundled ahead and back as instructed moving a total of 23 feet (7 m).

Pete Theisinger was speaking for a lot of us when he described himself afterward as "ecstatic." He told a reporter for *the Guardian* newspaper of Manchester, England, "We built a rover, so unless the rover roves we haven't accomplished anything." Amen to that. She had just proven that she *was* able to rove.

Just as the JPL teams had to negotiate with one another for time to do their testing in High Bay 1 and on the test beds, so with the scientists. Nobody would be able to get as much time as they wanted, especially in the early days, and there was always the fear no one wanted to talk about that Curiosity might have a fatal breakdown one day, leaving each team only with the data they had already collected. Their lists of experiments still to be run would have to wait many years for another ride to Mars, a dreaded possibility.

But as the day approached for John Grotzinger and his scientific teams to be given nearly carte blanche control of the rover, Curiosity was still performing brilliantly, passing all of the checkout tests without a hitch.

About a week or two before the much awaited sol when the scientists would be able to begin their work in earnest, a whole new style of daily operation swung into effect, following a routine invented for Spirit and Opportunity's rover operations that has become a standard.

The high-level decisions about which scientific activity and which instruments would be used by Curiosity for the next sol were put in the hands of what's called the Science Operations Working Group; in our shorthand, it's known as the SOWG (pronounced to rhyme with "log"). That group is made up of members of the scientific teams, elected by the teams themselves. Each instrument team submits to SOWG a detailed proposal of the next tasks they want to do with Curiosity, their best estimate of how long the task will take, and what the requirements are—for example, what time of day the work would need to be done, what areas of the nearby rock and soil they are interested in targeting, what tools, cameras, and instruments would be involved, and so forth. SOWG then integrates the requests.

These SOWG discussions go on in a room that reminds me of the General Assembly Hall of the United Nations, in miniature, with forty or fifty people in the room and a microphone for each SOWG member. But our meeting room has an added feature: It's also surrounded by MSL engineers sitting at computer terminals, providing minute-by-minute input on questions like how long it would take Curiosity to get from where it is now to a location suitable for the scientific work that this team or that one is asking to do. Or how much energy would be required for a proposed task. Or if there will be enough room in the computer memory to store the new data. (In time, most of the scientists would return to their home institutions, but would still take part in the SOWG activities and decisions by telephone, their voices being heard throughout the room as if they were present.)

As decisions are made for a particular set of activities, the engineers generate detailed sequence activity plans that will be the basis for what comes a bit later. This process has to happen quickly. If the SOWG doesn't get its job done in an hour or two before midnight on Mars, there might not be enough time to pull together the sequence scripts that the rover needs to receive just after it wakes up in the morning.

Sitting in on the SOWG discussions are some of the "Rover Planners," or RPs. They're also known as the "rover drivers," which makes it sound as if they steer, accelerate, and brake the rover moment by moment. They don't. The process is more complicated than that, as explained by one of the team, Vandana "Vandi" Tompkins. The daughter of a pilot with the Indian Air Force, Vandi spent her childhood moving with her family from one small town to another as her father was transferred. She earned two master's degrees, one in computer science, the other in robotics. In 1996, she read a newspaper article about plans to put the first rover on Mars: Pathfinder's little Sojourner. To her the idea was exciting enough that she set her sights on going to work at JPL, and she made it happen. She found herself "fascinated by the way scientists and engineers come together" to solve problems. Her easy and clear communication style and her technically astute detailed knowledge of the rover made her a perfect fit for the role of RP.

In the daily SOWG meetings, drivers listen to the negotiations among the scientists while gathering an understanding of what the goals and challenges will be in programming Curiosity for the day's work. Some sols the rover will be carrying out scientific tasks that involve hardly any moving. These are sols the rover can rest and recharging its batteries in preparation for the busy sols ahead. Other sols are designated for ordering the snail-paced rover to trundle along the terrain to a new location, covering as much as 400 feet (120 m) in one day. The drives involve a whole different set of procedures.

In laying the plans for a drive day, the rover drivers need to choose a route to cross landscape that is comparatively smooth, without any craters, and without terrain inclined more than about 15 or so degrees uphill or down. The drivers analyze the nearby terrain and specify waypoints along the preferred path using stereoscopic pairs of photos that they retrieve from Curiosity's cameras—sometimes driving the rover to a nearby higher elevation for a better view. They augment that data with images from the Mars Reconnaissance Orbiter spacecraft that circles the planet once a sol to come up with a long-term route.

Even with all the precautions, it's not possible from millions of miles away to choose a route guaranteed free of hazards, so Curiosity has been provided with backup protection. Thanks to the wonders of "machine vision" and modern robotics (and augmented by the wizardry of our brilliant software developers), when Curiosity is moving, she travels a short distance, then stops and takes stereo images of the terrain ahead. Her computer evaluates the images and determines if there are obstacles to avoid. If the path is clear, she navigates another short move. If she sees too many rocks or the rocks are too large, a steep hill, or some other hazard, she choses an alternate course. She also evaluates relatively better or worse areas to drive through and even remembers if she previously saw a particular obstacle, so she doesn't go in circles.

If Curiosity could speak for herself, these feats would definitely give her bragging rights.

One of the things Vandi finds especially intriguing is that when the scientists are pushing the limits of what's feasible, they need help from the drivers to find out what the rover can do, which, she says, "involves a lot of negotiating," Once the activities for the day have been set and approved, the planners have lots of questions they need answers to. For example, "Do you think the soil is cohesive here?" "What part of this rock do you want to explore?"

This information gathering needs to be thorough, but there's some urgency to the process. While the rover sleeps, the drivers have about six hours to complete their highly complex task, translating the plan from SOWG into a detailed list of step-by-step commands, then trying them out on a computer simulation, which displays how Curiosity will respond to each command.

Every sol, the drivers do a detailed walkthrough of every individual parameter and command. But that's not the end of the checking to guarantee accuracy. Vandi explains, "We always run a simulation in a visual rover planner simulation tool that runs an abstraction of the rover flight software and the rover. This is never skipped. We show the animation of this simulation to the entire room." Every item of the plan is scrutinized by every one of the planners on the team. Then they go through the entire sequence command by command with the entire room two more times.

When the drivers are finished creating the drive or arm instruction sequences, their work as well as sequences written by the science payload teams are turned over to the sequence integration team. The job of pulling together and seamlessly gluing together the sequenced commands is massive. Printed out, instructions for a single day can run to as much as fifty pages. Once the team is happy that the long list of commands does what they expected of it and will be safe, the command files are ready to be bundled up into a single file and sent on to Mars.

By this time it's early morning on Mars. The "alarm clock" has aroused the rover from its overnight sleep about an hour earlier, has already transmitted a ten- or fifteen-minute radio tone from its X-band high gain antenna to confirm that it is awake, is operating normally, and is ready to start the day's activities.

A little before 10:00 a.m. Mars time, one of the four or five people known as "Aces" comes on duty. (The nickname is actually the call sign used by members of this crew when they are setting up a radio communication.) The Ace is the interface between the project and the Deep Space Network.

This network has three transmission and receiving stations more or less equally spaced around the globe so that one of the three is always able to communicate in any required direction. Massive dish antennas capable of picking up the weakest spacecraft radio signals sent from hundreds of millions of miles from around the solar system are stationed at the Goldstone complex in the California desert near Barstow, in Australia outside Canberra, and in Spain not far from Madrid. Depending on the time of day, typically only one of these can see Mars at a time. The Ace works with the station operators to check that the appropriate dish antenna at the planned station is in position to transmit to Mars and to prepare the uplink signal for transmission at Curiosity's receiving frequencies. He or she then loads the file that contains all of the rover instructions, and hits the send button.

When the transmission is complete, the entire string of ones and zeros in the message spends the next ten to twenty minutes flying across the solar system. Once it arrives at Mars, Curiosity sends back a single X-band tone signal to confirm she has received the full message, then turns off her transmitter to save power, and begins to carry out the instructions for the upcoming sol. She is now ready for a full sol of exploration without any help from us. We won't hear back until late afternoon, after one of the orbiters flies overhead to collect Curiosity's loot of data.

CHAPTER 18
The Scientific Findings

While Curiosity was being thoroughly tested to confirm that she was healthy and fit for duty, the scientists were becoming anxious about getting started. It was two full weeks before the time came for testing the laser. The ChemCam instrument zapped a fist-sized rock selected as the target, given the name "Coronation," with thirty pulses, each delivering more than a million watts of power for a few one-billionths of a second.

Principal investigator Roger Wiens later commented, "The instrument worked almost too well: The signal was so strong it was slightly saturated on one of the peaks." But after shooting at another three rocks, he added, "The compositions [of the rocks] were not what we expected, and people began saying that the spectra didn't make sense. We began hearing, 'There's something wrong with your instrument.'"

In fact, there was nothing wrong with the instrument. "The volcanic rock chemistry of this area took us by surprise. It's different from any of the other landed places on Mars," Roger told me. "It was a real revelation."

About six weeks after Curiosity had landed, two of its instruments received their first trials when they were aimed at a fist-sized, pyramidal rock. The rock had been given the name "Jake Matijevic," in honor of MSL's Surface

Operations Systems chief engineer and former manager of the first Mars rover, the little Sojourner, who had passed away just days after the landing.

Like the Coronation results, these analyses showed this basaltic "lava" rock to be unlike any other ever found on Mars. It contained lower quantities of iron- and magnesium-rich minerals and far more silica-rich minerals than basalts that have been studied on Mars. This suggests that, just as we have discovered on Earth, Mars once had complex geologic processes that created a diversity of minerals in the bulk of the ancient bedrock. While it isn't yet clear what this means in the search for possible ancient life on Mars, it reinforces our belief that Mars is like Earth in many ways and shares some of Earth's diversity.

FIGURE 25. Pyramidal rock selected as the first target for APXS, MAHLI, and ChemCam laser trials. The ancient, wind-worn basalt revealed clues about the origin and diversity of the Martian interior. (Courtesy of NASA, JPL/Caltech, and Malin Space Science Systems)

Sometimes we search the world, and then make a discovery in our own back yard. That's something like what happened with Curiosity. In a distance from her landing spot no farther away than the length of one or two football fields, the science teams spotted what looked like concrete slabs sticking out of the ground. Embedded in the slabs were round rocks, looking like river rocks, or like the rocks you pick up on the shore to send skimming across the surface of the water.

What was so striking was that the rocks gave the appearance of having been cemented in place in the concrete slabs, and their shape clearly hinted that they had been eroded by rolling downhill, much like the rocks we find on Earth at the bottom of ancient riverbeds. This was our first glimpse ever on Mars of river rocks that had been eroded by water. From the orbiter image data, the scientists realized that the streambed Curiosity was driving over was probably far older than any streambeds on the surface of our own planet Earth.

It was one of those unexpected discoveries that can make science so exciting. The diverse members of the science community gathered at JPL were beside themselves with delight. It was thrilling just to observe their enthusiasm.

The brotherly spirit of the shared moment didn't last long, instead giving way to a big debate about where to go next. In the distance lay Mount Sharp, which had long been the focus, the place expected to produce the most valuable discoveries. Yet less than several days drive away from Curiosity's current location lay an area labeled Yellowknife Bay, a strange place where three kinds of terrain seemed to intersect, making it a tempting area to explore.

But this was in the wrong direction; it would mean going farther away from our route toward the base of Mount Sharp. In the end, the Yellowknife Bay advocates easily won the day because of the rich promise of its terrain.

As Curiosity approached Yellowknife, she was steered to a nearby small dune-shaped pile of sand and dust dubbed "Rocknest." There progress paused for a month while the science and JPL teams used the credit card–sized scoop on the end of the arm to gather a bit of that fine sand pile. After studying the sand in the scoop to ensure that it was free of pebbles, it was time to deliver the first sample of Mars terrain to the observation tray.

The examination gave the engineers and science team confidence that the sample would be safe for examination by SAM and the CheMin instrument. On a sol not long after, Curiosity picked up a scoop of sand and dust, ran it through the sieve, and finally deposited it into the CheMin instrument, which would obtain an x-ray diffraction pattern of the crystalline structure of the mineral grains.

I showed up in the CheMin room, where David Blake's science team had been waiting for the data. The place looked as if someone was having a birthday party. They were just analyzing the very first spectrum, and their instrument had revealed the crystallography of the minerals in the sample to a level of detail never seen before.

It wasn't radically different from what they had expected, but they could clearly see traces of three minerals well known to the scientists, if unfamiliar to most laymen: feldspar, pyroxenes, and olivine. The instrument had proven itself, working so well that the excited team members could only imagine what other details they might yet see in future samples.

Dave commented that this data was nearly twenty years in the making. Seeing real Mars data from his instrument was a dream come true. He and his team were grinning ear to ear, high-fiving one another. Grown men, all respected scientists with PhDs, acting like members of a college football team that had just unexpectedly defeated their archrivals.

The SAM team was just as excited when their instrument cooked up and smelled a bit of the Rocknest dust. They too were not overly surprised but they did detect the same odd and not-yet-understood perchlorate salts that Phoenix has detected in 2008. For the first time we had chemical analysis of the ubiquitous global Mars dust.

Sometimes even the most astute and intelligent among us fall into communications gaffes. One early conjecture based on data from Curiosity's SAM instrument turned out to prove embarrassing. The occasion was an interview by NPR reporter Joe Palca with our usually cautious John Grotzinger. Commenting on some of the early test results, John remarked, "This is one for the history books. The data is looking really good."

Palca, smelling the kind of major newsbreak that can make a reporter's reputation, later wrote, "Grotzinger can see the pained look on my face as I wait, hoping he'll tell me what the heck he's found."

It's part of the scientist's unwritten code that you don't announce what appears to be a breakthrough or important new finding until it has been published in a scientific journal. John tried to hedge with a reply meant to be noncommittal but that turned out to be easily misunderstood. "I know I'm killing you," he said.

What he meant was that the instruments were working as intended and Curiosity was sending back excellent data that the scientists were eagerly evaluating. What reporters thought he meant was, "We have found evidence of organic residue on Mars, but I'm not supposed to confirm it yet." The media went a little overboard reporting the news.

In a follow-up interview with NBC news reporter Mike Taibbi, John made a noble effort to correct the mistaken impression:

JOHN: "We're getting closer to understanding what it takes, how to explore systematically for organic compounds."

TAIBBI: "But not yet proof of life?"

JOHN: "Definitely not proof of life."

Next, in December, Curiosity was steered toward Yellowknife Bay. Once she arrived, it turned out she was surrounded by what appeared to be a bed of rock or clay—perhaps even the bottom of what had once been a small lake or standing body of water. But the scientists couldn't be sure what this stuff was until they analyzed a sample. The area proved to be rich with fascinating and strangely wind-carved features, including extrusions of sharp rock sticking out at unlikely angles. No one yet knew how long this area had been exposed or formed, but clearly something quite extraordinary had happened here long ago to create these superfine layers all around.

After spending a couple of weeks gathering scientific data along the rim of this shallow depression, Curiosity was moved back to a spot near were she had entered. The floor of this depression was not only flat, but it also looked much like flagstone pavement, a great spot to try the first drilling of a rock.

The development efforts of the SA/SPaH team were called into action. After the team's years of testing and struggle, their hardware and software was about to make it possible for humans to drill rock on another planet for the first time in history. The initial target was a rock given the name of the late John Klein, former MSL deputy project manager, a prince of a man who had been a great friend.

Preparations for drilling began early in the new year of 2013. Testing the drill with its hammer-action involved running through thousands of lines of computer code, much of that code dealing with the delicate process of moving the drill into the exact position and aiming it precisely, while monitoring the rate of motion and forces. As John Grotzinger vividly described it to *National Geographic*, they had to make sure the software would give the correct commands to "execute the hundreds of motions required to place a 65-pound drill as gently as a feather on a target the size of a pea."

The drilling would be straight down into the rock—first a trial divot hole to test the stability of the rock, then a deeper "mini-drill hole." This was a shallow test to make sure the "tailings" retrieved from rock would not show signs of deliquescence by turning into a paste.

Two days later, on February 8, sol 182, they finally took the plunge and bored a full, dime-sized hole about two inches (6 cm) deep. It was a landmark event, the first time ever that scientists had been able to capture a sample of Martian rock for analysis.

The drill tailings that came out of the red-colored rock were a grayish white. This was a good sign: it meant the site probably contained clay. But the truth needed to come from the SAM and CheMin instruments. Because of software missteps and motion-control tuning problems, it was nearly two weeks before a tiny portion of finely ground and filtered rock was dropped into SAM and CheMin for analysis.

It was another five suspenseful days before the on-board analysis was completed and the results sent back to Earth. Finally, in the morning hours of February 27, an email arrived telling us that the long-awaited SAM data had been received via the early-morning Odyssey orbiter relay pass.

Three hours later that same day, as I was checking my email over lunch, a particular message caught my eye. Mission manager Arthur Amador had forwarded an unusual communication, at the same time announcing the formation of a new anomaly investigation team. The message, from Johnny Greenblatt, an expert in avionic design who was the avionics subsystem operations lead for the day, alerted us to what we would come to call "the Sol 200 Anomaly."

Johnny's message said the data transmission that had just arrived from Curiosity contained "Warning" event reports: The flight software had

FIGURE 26. A self-portrait mosaic of Curiosity that combines dozens of exposures taken by the MAHLI camera. They were taken inside the spectacular Yellowknife Bay rock formation at the rover's first drill site (see Figure 27). (Courtesy of NASA, JPL/Caltech, and Malin Space Science Systems)

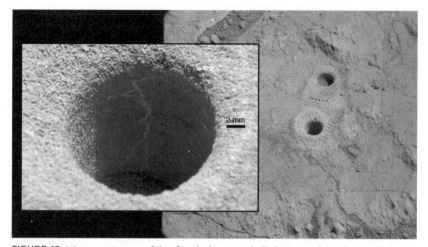

FIGURE 27. Mastcam image of the first holes ever drilled on another planet. Curiosity first drilled a shallow hole on the upper right to test the consistency of the rock, then a deeper hole (inset) to collect powdered rock for analysis by the on-board laboratories. (Courtesy of NASA, JPL/Caltech, and Malin Space Science Systems)

detected dozens of double bit errors (DBEs) in the computer's flash memory. At the same time there was another warning indicating that an attempt by the software to write files into the flash memory had failed.

Altogether, this status report on the condition of the rover's computer in English translated to something like, "I'm having a little trouble with my memory." It looked broken, to be sure.

Magdy Bareh, a gentle, thoughtful and determined engineer who is also an expert on the design of the computer and its software, was quickly assigned to be leader of the anomaly resolution team. He scheduled a 2:00 p.m. team meeting to figure out what might be going on. I wasn't going to miss this.

Less than an hour later, another transmission arrived that was even more chilling. Again, it was something we had never seen before. The rover had apparently ignored the command to put itself to sleep between the passes of the orbiters that relay its messages to Earth. What's more, even though the computer had remained awake, a few of the morning's planned tasks we expected it to complete were not getting done.

At the team meeting, more than a dozen people crowd into the small Fishbowl meeting room, so called because it has glass all around. For the time being, this would be our war room. The latest reports made it clear that the situation was even worse: The rover computer had ceased to obey and complete many of the commands in the list we had sent it after it had awakened this morning and on into the Mars afternoon. Yet the software barely complained. Normally, if there were something wrong with the computer, the software would halt and the backup computer would power up and take over. But instead the computer was still sending reports that meant, in effect, "I'm feeling fine, a little memory problem but otherwise nothing to worry about."

I couldn't help but think of the HAL computer in *2001: A Space Odyssey*, when it announced it would not obey what it was being told to do. Remember?

DAVE: Open the pod bay doors, HAL.

HAL: I'm sorry, Dave, I'm afraid I can't do that. This mission is too important for me to allow you to jeopardize it.

After a couple of hours of brainstorming and further study of the data, the team decided to see if we could reproduce the memory problem. Johnny and the lead fault protection engineer, Tracy Neilson, were given the job of seeing if they could figure out how to recreate this situation using the Mission Systems Test Bed. Tracy, an incredibly intelligent avant-garde engineer with a flare for detail, had played a key role in designing the rover system and its fault protection. She and Johnny set about experimenting to see if by corrupting the flash memory they might be able to reproduce these odd results.

Meanwhile, brilliant Ben Cichy, the software chief engineer, along with the developer of the software, the just as brilliant Rajeev Joshi, set up shop in the Fishbowl and began poring over millions of lines of code to see if they could find any clues about what was going on. So far, it looked as if the rover were still somewhat responsive to certain tasks it had been asked to do that morning—some tasks, but not all. In particular, any task that involved storing information to memory seemed to have stopped functioning.

I could see little beads of sweat on both their brows.

6:00 p.m. Magdy's anomaly resolution team was focused on trying to figure out how to come up with answers to the critical questions: Would a simple reboot of the computer fix the problem? Is it safe to just leave it alone? Or is this a problem we're just going to have to find a way to live with? The conclusions they were reaching were scary. The way the software was designed, it would be nearly impossible to come up with reliable answers to these questions.

Meanwhile, Johnny and Tracy called up our war room in the Fishbowl with a report: "We're able to reproduce the situation but you're not going to like it. The next time the rover tries to communicate, it will probably hang up and turn the radio off. The fault protection never trips and does not try to fix the problem."

We looked at each other. If that happened, we would never hear from the rover again, and would never be able to talk to it again. It would just sit there like a vegetable, permanently immobile.

That news left us sweating bullets.

1:00 a.m. at JPL—about 7:00 a.m. on Mars. They were still at it, and I was still trying to help Magdy with his go-forward plan. I was starting to lose my concentration, while his energy and focus was, as always, amazing. With the computer slowly going crazy, we had been throwing ideas for Johnny and Tracy to try out on the test bed. Time was growing short. Johnny and Tracy finally called, and their advice was, "It looks like we have no choice but to kill the computer—we need to force it into the isolation state"—a bit like a trance—"and make sure it stays that way." Magdy knew very well what this would mean; it was he who had designed in and tested this arcane method for dealing with a rogue computer. He reminded everyone that once we pulled the plug on the crazy computer and the backup computer mercifully took over, the faulty computer could do no harm, but we would be left without a second computer as backup. In the test bed, they had tried commanding the backup computer to take over, and it had worked.

So far, the main computer had refused to take naps or go to sleep for the night. This was good, as long as we could send commands to force the computer into the trancelike isolation state. But since the rover was supposed to be asleep at this hour, we didn't have any Deep Space Network stations scheduled for us to use.

Richard Cook called the Deep Space Network to declare a spacecraft emergency. This got us a station within the hour. With minutes to go, Magdy and the Ace sent the low-level commands to force the computer into its isolation state.

And then we each held our breath. Did our command get there in time? Will the backup computer wake up and take over?

Since Earth was now farther from Mars than when Curiosity landed in August, it now took twenty minutes for our commands to reach Mars, ten minutes for the computer to reboot, and another twenty for a report to come back.

3:00 a.m. at JPL. The moment was approaching when we hoped desperately we would receive word from Curiosity.

"Five . . . four . . . three . . . two . . . one . . ."

Nothing. No signal. We missed it.

Tracy knew something the rest of us didn't. Or maybe it was just that she wasn't willing to give up yet. She said, "Wait, wait." We fidgeted, but we waited. I started feeling weak in the knees.

Four minutes later, the signal showed up. It was from the B-side computer. Better late than never. We cheered. We realized later that this computer had not had its clock adjusted lately. The signal was late just because the computer's clock was slow—yet another bit of clock mischief.

It took three weeks to get the B-side computer fully configured and able to run the rover mission. We also were able to do some software workarounds on the broken A-side computer so if we ever had another problem that made us need to swap back to the A-side, it would be able to pick up its operations as it did during its healthy days, though now with only half as much flash memory available to it.

In one week more, Mars would be going behind the sun. For three weeks, our communications with the rover would be completely blocked. If we had not been able to fix the computer's problem before then and the same problem had recurred, Curiosity would have been dead by the time it was in position to communicate with it again.

We never were able to repair the broken flash memory, but we did figure out what was wrong with our software design. There should have been checks to make sure that if the main computer had a memory problem like this, or if it ever failed to go to sleep when asked, it would smoothly hand over its tasks to the backup computer. Discovering this led to some new software updates that we installed a few months later.

Still, today it makes me shudder to think how close we came to have the rover's mission end after so short a time.

While the engineering team and I worked feverishly to solve the Sol 200 Anomaly and ready Curiosity for "conjunction" (when Mars would go behind the sun), the science team was brimming with excitement. The CheMin and SAM data that arrived on the morning of sol 200 had confirmed that this flagstone-like pavement Curiosity was sitting on was a type of superfine layered material called phyllosilicate clay—layers of clay so fine that it resembled phyllo dough pastry.

What's so exciting about this? On Earth, this type of clay is found where slow-moving and long-lasting water has laid fine sediments layer by layer over time. Clays chemically preserve a record of the climate at the time they were created, billions of years ago. Even more promising, it is also in clays that we find chemical records of ancient life. Clays have the potential of preserving organics for many hundreds of millions of years.

Further analysis by SAM showed that this clay was of a type that can only form in fresh water with neutral pH conditions—the conditions like our own lakes and oceans that are breeding grounds for life.

Was this sufficient proof that Mars was once habitable for single-cell organisms? The samples showed that Mars had been wet, it had been warm, it had had a neutral pH, it had not been too salty, and the water had been present for a long time. The samples also contained all of the elements that life needs to survive, the so-called CHNOPS atoms: carbon, hydrogen, nitrogen, oxygen, phosphorus, and sulfur.

These were all the ingredients that would be needed. But could life survive and *thrive* in this rock? To thrive, any Mars organism would also need an energy source: food.

On Earth we have microorganisms called chemolithotrophs. These organisms derive energy from chemical reactions in their single-cell bodies that oxidize chemicals they get by "eating" the rock. Imagine: a living organism that survives by ingesting rock! Long ago on Earth this was the norm and not the exception. The conditions found at the drill site plus some sort of chemical energy source would be all that these types of organisms would need to survive and thrive. If the scientists found a chemical energy source, they would be able to confirm that these rocks were once habitable for this kind of organism. In fact, they *did* find a chemical energy source. The drilled sample "John Klein" contained sulfates and sulfides that simple chemolithotrophs should thrive on. Everything was there.

Beyond anyone's expectations, the results of testing the material from this first rock ever drilled on Mars produced evidence that conditions on ancient Mars had indeed been suitable to support a thriving microbial community had it been there. To scientists worldwide, the evidence that Mars had once been capable of supporting life was spectacular.

About 3.5 billion years ago, a large freshwater lake right at Curiosity's landing site could have had microbial life—around the same time that life may have been beginning on Earth. These were electrifying findings, a scientific bonanza.

John Grotzinger told a press conference, "We have found a habitable environment that is so benign and supportive of life, probably if this water was around and you had been on the planet, you would have been able to drink it." He also said, "All the essential ingredients for life were present," a scientifically stunning statement. He called this "a critical turning point in the mission."

The very first rock drilled had provided the answer to the number-one goal of the entire Mars Science Laboratory mission. No matter what else Curiosity may reveal during its lifetime on Mars, this will inevitably be looked on as the major achievement of the mission.

There was more. Analysis of the Yellowknife rock brought surprises as well. Rocks exposed on a surface are bombarded with cosmic rays, causing the formation of the gases neon-21, helium-3, and argon-36 to accumulate in the rock. Over a long period of time, any traces of organic matter that might have existed near the surface are destroyed.

If Gale Crater's surface had been exposed to cosmic rays over a long enough time, traces of organic matter anywhere within reach of Curiosity's drill would have been destroyed. There would have been no possibility of finding whether life had ever existed at Gale.

But in the Yellowknife rock, the quantities of these gases were smaller than the scientists had expected. The data suggesting that the surface of Gale may have been exposed to cosmic rays for only something like 80 million years—in geologic terms, just yesterday.

This finding brought smiles from the scientists. The news of this short exposure holds out hope that Curiosity may find traces of organics stored in the protective layers of clay that surround the base of Gale Crater's mountain, Mount Sharp.

On June 6, 2013, six months after Curiosity arrived, she set out on her cross-country trip heading for Mount Sharp, a voyage of about 5 miles (8 km) that could take up to a year, which seems a long time as I write. But assuming all goes well, she will have arrived by the time you read this. We have every expectation that there will be other major scientific discoveries to add to the ones recounted here.

At Mount Sharp, the scientists are hoping to find clues explaining what caused this high mountain to form as a massive layer cake. If the soil was deposited in layers, what were the influences that carved them out, exposing them for us to see from space? The answers may hold clues to whether we will find organic residue, signs of ancient life, at this mysterious mountain.

When we began to work on what came to be called the Mars Science Laboratory, with its rover Curiosity, we could hardly have begun to guess how rich a contribution this mission would make to our knowledge about the ancient history of Mars and, along the way, the ancient history of our own planet Earth.

Curiosity has revolutionized the art of planetary exploration. An entire area of scientific research has been opened. Scientists now can take a ride as remote members of an expedition to a planet. They can make decisions of where to explore and what samples to examine, much like scientists on a field trip on Earth. It's as if they were sitting inside the rover, steering, driving, operating the drill and the scoop, depositing samples into the instruments, then patiently waiting for the results.

Moreover, Curiosity provides an unprecedented breadth of extraterrestrial exploration that can now span many orders of magnitude in scale, from the 100-kilometer scale of Mount Sharp down to the submillimeter scale of a mineral grain, and even smaller, to the scale of complex organic molecules and their constituent atoms.

We had created a new generation of extraterrestrial exploration.

* * *

FIGURE 28. Sparkling in brilliant sunshine, Curiosity looks back at her own tracks as she slowly makes her way to the base of Mount Sharp. (Courtesy of NASA, JPL/Caltech, and Malin Space Science Systems)

While Curiosity continues to produce a wealth of intriguing new data, I have moved on, diving into my next JPL project. The springboard came from the Decadal (once every decade) Survey conducted at the request of NASA and the National Science Foundation. The 2013 Survey, chaired by Professor Steve Squyres of Cornell University, provided a group of "flagship priorities." Based on input from a large number of scientists, Squyres and his survey team chose a Mars project as the number-one flagship priority.

It was announced with these words: "The view expressed by the Mars community is that Mars science has reached a point where the most fundamental advances will come from study of returned samples." In line with that, the next step in Mars exploration is sample return, aimed at bringing back half a kilogram of carefully selected Martian rock cores. The sample return project begins with another version of the Mars Science Laboratory, but this one with an added goal: "sample caching," meaning collecting and storing rock core samples from Mars for eventual return to Earth, to be handed over to the world's leading science laboratories for analysis.

Some of the hurdles we're facing will sound familiar, including a new type of still-larger supersonic parachute. The target launch date for the sample collecting and caching rover is 2020. Beyond that, we will need larger landers that can retrieve the cached samples and rocket them into Mars orbit for capture and return to Earth. I am now focusing on the design and testing of new ways to slow these larger spacecraft for landing. Some of the techniques we are testing might well be used to slow and safely land the first human explorers on Mars.

Meanwhile, as of this writing, valuable scientific data is continuing to flood in from Curiosity. ChemCam alone has produced more than 100,000 laser firings, producing data from each, available to any interested scientist. As of fall 2013, hundreds of scientific articles based on Curiosity's data had already been published, with new articles being published virtually every day. Even so, PI Roger Wiens says, "I've come to realize that the most interesting discoveries MSL could make are the ones that we have no idea about yet."

Though a grueling challenge, in the end this Mars Science Lab project turned out to be an extremely gratifying adventure. I would do it all again in a heartbeat. Second only to my wife and child, the opportunity of working on space exploration is the greatest gift life could have offered me.

FIGURE 29. Curiosity's view of her destination, Mount Sharp. The layers of terrain, suggestive of the Grand Canyon, offer the possibility that within them are clues that might open a new chapter in our understanding of the story of life in our solar system. (Courtesy of NASA, JPL/Caltech, and Malin Space Science Systems)

Appendix—NASA's Mars Missions

NAME	LAUNCH DATE	RESULT	DETAILS
Mariner 3	1964	Failure	Shroud failed to jettison
Mariner 4	1964	Success	Returned 21 images
Mariner 6	1969	Success	Returned 75 images
Mariner 7	1969	Success	Returned 126 images
Mariner 8	1971	Failure	Launch failure
Mariner 9	1971	Success	Returned 7,329 images
Viking 1 Orbiter/Lander	1975	Success	Located landing site for Lander and first successful landing on Mars
Viking 2 Orbiter/Lander	1975	Success	Returned 16,000 images and extensive atmospheric data and soil experiment data
Mars Observer	1992	Failure	Lost prior to Mars arrival
Mars Global Surveyor	1996	Success	Mapped Mars. Discovered layered terrain as well as recent water activity
Mars Pathfinder	1996	Success	Technology experiment lasting 5 times longer than planned for
Mars Climate Orbiter	1998	Failure	Lost on arrival

(continued)

NAME	LAUNCH DATE	RESULT	DETAILS
Mars Polar Lander	1999	Failure	Lost on arrival
Deep Space 2 Probes (2)	1999	Failure	Lost on arrival (carried on Mars Polar Lander)
Mars Odyssey	2001	Success	High resolution images of Mars
Mars Exploration Rover—Spirit	2003	Success	Operating lifetime of 2210 days working on Mars, more than 24 times design life
Mars Exploration Rover—Opportunity	2003	Success	At the time of publication still operating after more than 3675 days working on Mars, more than 40 times design life
Mars Reconnaissance Orbiter	2005	Success	Returned more than 26 terabits of data (more than all other Mars missions combined)
Phoenix Mars Lander	2007	Success	Returned more than 25 gigabits of data
Mars Science Laboratory	2011	Success	Now exploring the habitability of Mars
Mars Atmosphere and Volatile Evolution	2013	En route	On the way to Mars

Source: NASA

Acknowledgments

The person most responsible for encouraging me to take risks throughout my long career and for helping me be the best I can be is my wonderful wife, Dominique. She has quietly suffered long hours of my absences, both physically and mentally, while also being a sounding board and coach who reminded me that I could do it. She is the heart behind my absentminded brain. Caline, my lovely daughter, I too must acknowledge and apologize for the many weekends and lost father time that this manic and obsessive-compulsive work forever stole from her. I love you both dearly. Caline, I fervently hope that you too will discover your passions, and your own wonders of the universe.

When my brilliant friend Brian Muirhead suggested that I take my threats to write a book seriously, he connected me with my talented coauthor Bill Simon. Bill took a huge gamble when he agreed to work with me on this book project. He quickly learned that my family and my Mars work come first. Nevertheless, like the MSL team, he persevered for a long, difficult year. This is his book as much as mine.

Although my colleagues and I try to conquer bad luck with mind-numbing checking, double checking, and testing of every detail while bombarding ourselves with myriad questions about what could go wrong, most of our luck is wrapped up in the genuine talent and dedication and perseverance that so many people have brought to the success of MSL as well as the success of the entire Mars exploration renaissance that has occurred in the past fifteen years. We are truly and unquestionably lucky to have access to the vast diversity of talent, integrity, and skills that my teammates from around the country and beyond have demonstrated again and again. I stand amazed that people like this even exist on our little planet, let alone that I have been so privileged to walk among them. How did I get so lucky? When the wheels of the last rover stop turning years from now, we all should be proud to know that the tracks were laid by the toil and talent of the best of our humanity. To my Mom, for her endless joy and wonder at life, my Dad for being my biggest fan, my brother Jack, and my extraordinary identical twin Chuck (who for some freak of nature got all of the real talent between us), I must say that I love you and that you should be tired of me by now!

I'm glad to thank as well the US taxpayer for allowing us to do these odd things. I truly believe that we visit the planet Mars in order to take another step toward greater wisdom and to extend ourselves beyond our own shrinking yet precious world. I hope that these robotic missions to Mars will renew in all of us an invigorated sense of humility and awe about our place in this wonderful universe we are so privileged to roam.

May humanity's children experience the wonder of these discoveries and their own for many generations to come.

—Rob Manning

This book came about thanks to JPL whiz kid Brian Muirhead, with whom I had done an earlier book, *High Velocity Leadership*. I will forever be in Brian's debt for putting me together with the amazing and inspiring Rob Manning.

How can you not admire a man who sends his spacecraft off on the cross-country journey to the launch pad by hauling out his trumpet and playing *When the Saints Go Marching* as the trucks drive away?

Rob and I are fortunate to have Carolyn Gleason and her talented crew at Smithsonian Books as our publisher. The guidance of a wise, insightful, supportive editor is what every author dreams of, and Carolyn fills the bill.

I can't offer enough praise, as usual, to Bill Gladstone, the founder and head of Waterside Productions, and my agent through the whole of my thirty-book writing career.

I owe a debt to the insightful Elisabeth Sedano, whose comments on the manuscript were highly valuable, and a special tip of the hat to a couple of people who went out of their way to provide input and respond to questions: John Grotzinger, and rover planner (a.k.a. "driver") Vandi Tompkins.

Finally, my great appreciation and admiration for the people closest to me, who put up with my too pervasive distance and lack of awareness as I devoted unconscionable amounts of time to this fascinating but draining project. To my precious Victoria, to David and Sheila, Lisa, Abel, and Susy—thanks for your support and for being part of my life. And to my darling Charlotte: You make it all worthwhile.

—Bill Simon

Index

Illustrations are indicated in *italics*.

A

actuator
  cold, 124–25
  heaters, 126
  high-gain antenna, 75
Adler, Mark (rover mission manager), 32, 42–43
aeroshell, 60, *60*, 143–44, 182–83
airbag landers, 11, 22, *23*, 24–25, 32, 62
Alpha-Particle-X-ray-Spectrometer (APXS), *76*, 77, *95*, 132, *198*
APXS. *See* Alpha-Particle-X-ray-Spectrometer
Arcjet test facility, 106
Armstrong, Neil (Apollo astronaut), 21–22, 50
assembly, test, and launch operations (ATLO)
  activities, 117
  clean room, 116, 118
  control room, 152
  crew at Cape Canaveral, 160
  flight equipment, 140
  High Bay, 117, 138
  manager, 135
  schedule, 135
  team, 117–18, 130, 133, 134, 135, 144, 147, 151, 157
  test conductors, 151
avionics (electronics and cabling), 92, 141, 148, 149, 163, 202

B

backshell, 36, 60, *60*, 185–86
bacteria
  Earth, 4–5, 116
  Martian, 12
Bareh, Magdy (computer software engineer), 204–5
Blake, David (NASA Ames Research Center), 77, 200
Braun, Bobby (entry-system aerodynamics engineer), 47, 101–2

bridle and umbilical device, *60*, 61, 146
bridle ropes, *31*, 42, 133, 146
  Vectran, 186
bubbleheads, 25, 27, 31
bunny suit, 116, 147

C

CAB. *See* Centro de Astrobiología
Cape Canaveral
  Mars Science Laboratory, 134, 158, 160, 162
  Spirit and Opportunity mission, 39
carbon-dioxide molecules, 4
  Martian CO_2 dust, 16
  Martian CO_2-laden atmosphere, 184
ChemCam. *See* Chemistry and Camera
Chemical/Mineralogical X-Ray Diffraction Instrument (CheMin), *76*, *77*, *95*, 132, 199–200, 202, 208
Chemistry and Camera (ChemCam), *76*, 79, 81–86, *98*, 104, 197, *198*, 212
chemolithotrophs, 208–9
CHIMRA. *See* Martian Rock Analysis instrument
chips, superhigh-density programmable, 148
CHNOPS atoms, 208
Climate Orbiter. *See* Mars Climate Orbiter
cold actuator, 124–25
cold chamber, 124
computers and computing
  architectures, 11
  backup, 142
  cards, flight, 165
  chip, 166
  circuits, 166, 177
  controlling the landing, 38
  for firing thrusters, on-board, 22
  lander, 62
  memory, 11, 57–58
  models, 91
  network wiring, 134

on-board, 145
rendering of the sampling process, 94
rover's, 36, 59, 62, 65, 122, 142, 159, 162,
    165–66, 173–74, 177, 181–83, 186, 190
simulations, xiv, 53, 61–62, 100, 101,
    146, 154, 166, 171, 173–76, 197, 194
spacecraft, xiii, 11
test bed, 60
computer code, xii, 41, 43, 131, 149, 152,
    157, 190, 202, 205
Cook, Richard (flight systems manager),
    43, 55, 90, 93, 107, 115–16, 140, 207
cruise propulsion fuel lines, 171
cruise stage, xiv, 16, 60, *60*, 108, 114, 117,
    144, 158, 170, 173–75, 182
cruise stage propulsion system, 174
cruise stage separation, *183*
Cruz, Juan (NASA Langley Space Cen-
    ter), 52–54
Curiosity rover (2011). *See also* Mars Sci-
    ence Laboratory (MSL), xiv, 35, 149,
    150, 157–58, 159–61, 162, 165–66,
    169, 182–88, *183*, *184*, *185*, *186*, *187*,
    188, *189*, 201, 202, 204, 207, 210
autonomous entry guidance system, *155*
coded message beamed back to Earth, xii
cost, xv, 190
descent to Martian surface in Gale Cra-
    ter, 215
EDL sequence of events, 182–88, *183*,
    184, *185*, *186*, *187*, *188*
fault protection, radical redesign of rov-
    er's, 142
fully built rover, *150*
landing sites
choosing, 153–57
Gale Crater, xv
Mount Sharp landing spot, base of, 156–
    57, 183, 189, *189*, 199, 210, *211*, *212*
nuclear power source, 142
pictures of, *x*, *203*, *211*
rover separates from descent stage, 133–34
"sample playground" added, 141
test driving, 151–52, *152*
wake/sleep schedule, 121–23

D

D'Amario, Lou (navigation team chief),
    170, 173

DAN. *See* Dynamic Albedo of Neutrons
DBEs. *See* double bit errors
Decadal Survey, 211
decelerator, supersonic inflatable aerody-
    namic, 53
decelerator, supersonic retropulsive, 69
Deep Space 2 Probes (1999), 215
Deep Space Network, 170–71, 173, 195, 206
descent engine throttle motors, 157
descent engines, 24, *60*, *186*, 190
descent stage, 60–63, *60*, 70, 117, 133–34,
    143–46, 158, 170–71, 177, 185–86,
    187, 188
Donaldson, Jim (avionics expert), 147–
    48, 163–64
double bit errors (DBEs), 204
drill bit(s)
about, *95*, 96, 145, 160–61
mission reclassified by the Planetary
    Protection office, 161
spare, *95*, 141, 145, 160
Dynamic Albedo of Neutrons (DAN), *76*, 78

E

Earth landers, 20
Edgett, Kenneth S. (Malin Space Science
    Systems), 77, 86, 88, 132
EDL. *See* entry, descent, and landing
Elachi, Charles (JPL director), 85–86,
    136, 176–77
electrical interference, 62
electronics test bed, 109
English-to-metric error, 14
entry, descent, and landing (EDL)
about, 11, 29
airbags, 11
automated, 51
Curiosity's EDL sequence of events,
    182–88, *183*, *184*, *185*, *186*, *187*, *188*
design "taxonomy" chart of the EDL
    design, 27, 29, *29*
heat shield, 11
solid rockets, 11
supersonic parachute, 11, 27, 52, 59, 69,
    97, 184, *185*, 212
test bed, 39–40, 60, 109, 116–17, 123, 144,
    148, 161–65, 170–73, 176–78, 191, 205–6
entry guidance, 22

Exploration Rover. *See* Mars Science Lab-
    oratory (MSL); Spirit and Opportu-
    nity missions

F

F-18 fighter jet, 145
fabrication facility, 117
fault protection, 141–42, 144, 151, 205–6
Figueroa, Orlando (CDR panel head), 101–2
flash memory, 108, 204–5, 207
flight unit, 117, 144
fuel leak simulation, 173–75
fuel tanks, 24, 27

G

Galileo spacecraft, 10–11
Gas Chromatograph (GC), 76
Gavin, Tom (flight projects manager), 113, 136
GC. *See* Gas Chromatograph (GC)
Gellert, Ralf (Max-Planck-Institute for
    Chemistry), 77, 132
Genesis mission, 82
Global Surveyor. *See* Mars Global Surveyor
Goddard Space Flight Center, 10, 76, 126
Greenblatt, Johnny (avionics expert), 202,
    205–6
Griffin, Mike (NASA administrator), 66–
    71, 99, 115, 139
Grotzinger, John (project scientist), 103–
    4, 145, 156, *156*, 191, 200–202, 209
Gruel, Dave (ATLO manager), 135, 160

H

Hand Lens Imager. *See* Mars Hand Lens
    Imager (MAHLI)
Hassler, Donald (Southwest Research
    Institute), 77–78
hazard sensor, 57
hazcams, 189
heat shield
    about, 60, *60*, 106–7, 158, 182–85
    for Earth landings, 20–21
    for Mars landings, 20
    separation, 79
Hubble Space Telescope, 137
Human Planetary Landing Systems panel
    (2004), 46–55
human-scale landers, 51, 53
hydrothermal vents, 4

hypersonic aircraft, 21
hypersonic inflatable aerodynamic decel-
    erator, 53

I

Instrument payload, 59

J

James Webb Space Telescope (JWST), 137
Jet Propulsion Laboratory (JPL) (Pasadena)
    Cruise Mission Support Area, xi, 41, 165
    High Bay, 116, 138, 144, 147–48, 157–58, 191
    test bed, 39–40, 60, 109, 116–17, 123, 144,
        148, 161–65, 170–73, 176–78, 191, 205–6
Johnson Space Center (Houston), 10, 21, 51
Jolly, Steve (Lockheed), 29–30
JWST. *See* James Webb Space Telescope

K

Krajewski, Joel (systems engineering), 31, 141

L

lander(s)
    airbag, 11, 22, *23*, 24–25, 32, 62
    computer, 62
    Earth, 20
    human-scale, 51, 53
    legged, 22, *23*, 62, 65
    pallet, 27, *28*, 32, 37
    rover-delivery, 24
    software, 24
    three-legged, 24, 56
landing ellipses, 21, 35, 155, *155*, 157, 184
landing sites on Mars, 153–57, *155*, *156*
launch–abort mode, supersonic, 49
Lee, Gentry (JPL engineer), 112–13, 119,
    134, 136
Lee, Wayne (EDL chief engineer), 13, 41–42
legged landers, 22, *23*, 62, 65. *See also*
    three-legged lander
Li, Fuk (JPL Mars program manager), 67,
    70–71, 107, 113, 119
Lockheed Martin, 106, 119
Lockheed Martin Astronautics, 15, 158

M

MAHLI. *See* Mars Hand Lens Imager
Malin, Mike (Malin Space Science Sys-
    tems), 75, 78–79, 86, 88

MARDI. *See* Mars Descent Imager
Mariner spacecraft, 214
Mars
  astronauts on long or short stays, 48
  basaltic "lava" rock, 198
  CHNOPS atoms in all samples, 208
  concrete slabs sticking out of the ground, 199
  crystallography of minerals show feld-
    spar, pyroxenes, and olivine, 200
  Eberswalde, Gale, Holden, and Mawrth
    landing sites, 154
  formation of, 4
  Gale Crater landing site, xv, *155*, 156–57,
    170, 174–76, 182–83, *188*, 191, 209–10
  human-scale landers, 51, 53
  "Jake Matijevic," pyramidal rock, 197–98, *198*
  "John Klein" drill sample contained sul-
    fates and sulfides, 209
  landing ellipses, 21, 35, 155, *155*, 157, 184
  landing hazards, 20, 27
  landing sites, choosing, 153–57
  lower gravity and different environment, 61
  Melas Chasm, 154
  meteorite (found in Antarctica), 12
  Mount Sharp, 87, 156–57, 183, 189,
    189, 199, 210, *211*, *212*
  organic residue, evidence of, 201
  perchlorate salts, 200
  perchlorate in the soil, 121
  phyllosilicate clay, 208
  reservoir of ice in northern plains, 121
  "Rocknest," dune-shaped pile of sand
    and dust, 199, 200
  rocks possibly showing deliquescence,
    141–42
  "sample return" space mission, 12
  soil with consistency of thick mud, 132
  spacecraft sent to, xii–xiii
  thermal infrared observations of sand
    dunes, 87
  wind erosion and transport of sediment, 87
  wind gusts or dust storms, 58, 66, 132
  Yellowknife area and rock analysis, 209
  Yellowknife Bay, 199, 201, *203*
Mars, early theories and stories about, 1,
    8, 9, 20–21, 25
Mars Descent Imager (MARDI), *76*, 79, 86, *98*
Mars Exploration Rover—Opportunity
    (2003), *23*, 34–41, 43–44, 46–47, 52, 55,
    57–58, 66–67, 82, 90–91, 101, 110, 121,
    130–31, 153, 163–65, 173, 191, 215
Mars Exploration Rover—Spirit (2003),
    23, 34–41, 43–44, 46–47, 52, 55, 57–
    58, 66–67, 82, 90–91, 101, 110, 121,
    130–31, 153, 163–65, 173, 191, 215
Mars Exploratory Rover (MER), 34. *See
    also* Mars Science Laboratory (MSL)
  design, 32–33, 36–37
  EDL software determined when rover
    was on Mars, 62
  landing systems, 37, 62
  launch date, 36
  rover-on-a-rope, 37
Mars Global Surveyor (1996), 24, 57, 87,
    97, 163, 214
Mars Hand Lens Imager (MAHLI), *76*,
    77, 86, *95*, 142, *198*, *203*
Mars Mobile Pathfinder, 33–34
Mars Observer (1992), 214
Mars Odyssey (2001), 202, 215
"Mars Options Assessment Review," 32
Mars Orbiter Camera, 97
Mars Pathfinder (1997), 12, 24–25,
    27–28, 42, 89
Mars Polar Lander (1999), 15, 32, 215
Mars Reconnaissance Orbiter (2005), 34,
    119, 156–57, 170, *188*, 193, 215. *See
    also* Spirit and Opportunity missions
Mars Science Laboratory (MSL). *See also*
    Curiosity rover; Mars Exploratory
    Rover (MER), 58–59, 60, 66, 73–74,
    90–93, 100–102, 109, 115–16
  actuator with electric heater strips
    wrapped around them, 125
  bridle ropes configuration, *31*
  camera locations, *98*, 99
  design error risk, 66
  heat shield, 106
  heaters for motors and gearboxes, 125
  instrument list, 75–79, *76*, 88
  instrument payload, 59
  major components of MSL, *60*
  "Mars Smart Lander" rechristened
    "Mars Science Laboratory," 58
  mockup of the rover, full-scale wooden, 93
  NASA's scientific and instrumentation
    goals for MSL, 74
  pallet lander concept, *28*

Mars Science Laboratory (*continued*)
 sample-handling process, 95
 science instruments, locations of ten, 76
 single-string to redundancy approach, 93
"Mars Smart Lander," 34–35, 58. *See also*
 Mars Science Laboratory (MSL)
 team, 35–36
Martian Rock Analysis instrument
 (CHIMRA), 96, 101, 125, 141
Mast Camera (MastCam), 76, 78, 86, 98, 204
Matijevic, Jake (MSL's Surface Operations
 Systems chief engineer), 197–98
McCuistion, Doug (Mars Exploration
 Program director), 75, 136, 139
mechanically deployable entry system, 53
"MegaRover," 34. *See also* "Mars Smart
 Lander"
MER. *See* Mars Exploratory Rover
Meyer, Mike (Mars program scientist),
 75, 84–85, 104
Mission Systems Test Bed, 117, 159, 161,
 171, 205
molybdenum disulfide (gear lubricant),
 124–25
moon landing, first, 20
Moreno, Victor (electrical systems engi-
 neer), 133–34
motors and gears, titanium, 124–25
MSL. *See* Mars Science Laboratory
Muirhead, Brian (JPL spacecraft manager), 11, 25
multicellular organisms, first complex, 5

N

NASA
 Ames Research Center (Mountain
 View, California), 51, 97, 107
 animated video of MSL landing on Mars, xii
 Decadal Survey, 211
 Langley Space Center (Virginia), 52
 National Aeronautics and Space Act, 10
 Orion program, 107
 Science Mission Directorate, 99
 small thrusters to stabilize capsule in
 outer space, 21
National Aeronautics and Space Act (1958), 10
National Full-Scale Aerodynamics Com-
 plex (NFAC), 97
Navcams (cameras), 98

NFAC. *See* National Full-Scale Aerodynamics Complex (NFAC)
nuclear power source, 142

O

Observer. *See* Mars Observer
Odyssey. *See* Mars Odyssey
Operations Readiness Tests (ORTs), 170–78
Orbiter. *See* Mars Reconnaissance Orbiter
Orbiter Camera. *See* Mars Orbiter Camera
Organic Check Material, 95
ORTs. *See* Operations Readiness Tests

P

pallet lander, 27, 28, 32, 37
parachute(s)
 clustered, 27
 component of MSL, 60
 Kevlar ribbing, 98
 multistaged, 53
 of nylon fabric with Kevlar suspension
 lines, 96–97
 Phoenix spacecraft during landing, 119,
 120, 121
 subsonic, 36, 59
 supersonic, 11, 27, 52, 59, 69, 97–98,
 184, 185, 212
 testing, 97–98, 118
Pathfinder. *See* Mars Mobile Pathfinder;
 Mars Pathfinder
pendulum dynamics, 70
PFRs. *See* problem/failure reports
phenolicimpregnated ceramic ablator
 (PICA), 107
Phoenix Mars Lander (2007), 105, 215
 Discovery of soil with consistency of
 thick mud, 132
 legged lander, 23
 on its parachute during landing, 119, 120, 121
PICA. *See* phenolicimpregnated ceramic
 ablator
planetary exploration missions, 122
Polar Lander, 15, 25
portioner (apportions filtered material),
 132, 141
Powell, Dick (NASA Langley), 47, 49
power generators, 59, 126
power plants, radioactive, 48
problem/failure reports (PFRs), 129

prokaryotes, 4
propulsion design, precision-controlled, 29
propulsion system
    cruise stage, 174
    landing, 27
    plumbing diagram, 56
prototype, 100, 117
pulsed engines (Polar Lander), 28
pyrotechnics, 62, 164

Q

Quadruple Mass Spectrometer (QMS), 76

R

RAD. *See* Radiation Assessment Detector
radar
    about, *60*, 145, 154, 157, 185–86
    altimeter, 15
    imaging, 36
Radiation Assessment Detector (RAD),
    76, 77–78
radio antenna, 60–61
radioisotope thermoelectric generator
    (RTG), 122
radio spectrum analyzer, 16
Rasmussen, Bob (guidance-and-control
    engineer), 37–38
Ratliff, Martin (space radiation expert), 172–73
Reconnaissance Orbiter, 156–57, 170, *188*,
    193, 215. *See also* Mars Reconnaissance
    Orbiter; Spirit and Opportunity missions
Redundant components and systems, 92,
    142, 178
REMS. *See* Rover Environmental Moni-
    toring Station
retropropulsion, supersonic, 53
retropulsive decelerators, supersonic, 69
RIMU. *See* Rover Inertial Measurement Unit
Rivellini, Tom (airbag designer), 27, *27*,
    32, 37, 38
robotic
    arms, 59, 92, 95, *95*, 96, 102, 124, 131, 139
    Mars mission to test new technologies, 54
rock brush, *95*
rock from meteor impacts, 5
rocket plume damage, 38
rover-delivery lander, 24
Rover Environmental Monitoring Station
    (REMS), *76*, 78, 190

Rover Inertial Measurement Unit
    (RIMIU), xiv
rovers, nonredundant, 59
RTG. *See* radioisotope thermoelectric
    generator

S

SA. *See* sample arm
Sabahi, Dara (JPL's chief mechanical
    engineer), 29–30, 32, 105–6, 109–110
SAM. *See* Sample Analysis at Mars
Sample Analysis at Mars (SAM), 76, *76*, 77,
    95, 126, 132, 160, 199–200, 202, 208
sample arm (SA), 93
Sample Arm/Sample Processing and
    Handling subsystem (SA/SPaH), 93–
    94, 96, 101, 107, 160, 201
sample-handling equipment, 58
San Martin, Miguel (guidance-and-control
    engineer), 30–32, *31*, 36, 43, 57, 63, 147
SA/SPaH. *See* Sample Arm/Sample Pro-
    cessing and Handling subsystem
    (SA/SPaH)
Schmitt, Harrison "Jack" (Apollo 17
    astronaut), 46, 48–49
Schratz, Brian (EDL communication
    team), xiv, 170
Science Operations Working Group
    (SOWG), 191–93
simulations, computer, xiv, 53, 61–62, 100,
    101, 146, 154, 166, 171, 173–76, 179, 194
single-cell life, 5
sky crane
    landing phase, testing of, 145–46
    maneuver, 57, 61–62, 69, 133, 177, 186, *187*
    phase, 61, 63, 133
SLA. *See* Super Lightweight Ablator
software
    code from the Spirit and Opportunity, 131
    created for every part of the EDL oper-
        ation, 131
    cruise, 131
    to guide the robotic arm, 131
    lander, 24
    Mars entry guidance, 22
    test bed, 163–64
solar array power supply, 48, 121
solar flare, 173
solid rockets, 11, 25, 27–28

SOWG. *See* Science Operations Working Group

Space Shuttle, 20, 46–47, 49–50

Space-X team, Elon Musk's, 69

SPaH (smaller arm), 94

spare drill bits, *95*, 141, 145, 160

Spirit and Opportunity missions, 23, 34–41, 43–44, 46–47, 52, 55, 57–58, 66–67, 82, 90–91, 101, 110, 121, 130–31, 153, 163–65, 173, 191, 215

  airbag lander, artist's rendering of, *23*

  backup antenna, 43

  blown fuse, 40–41

  budget, 90

  computer code fixes, 40–41

  landings, 42–43

  launch preparations, 38

  untethered roving capability, 42

star-sensing equipment, 60

Steltzner, Adam (EDL team head), xiv, 67, 69, 106–7, 118, 170, 175–76

Stern, Alan (NASA associate administrator), 99, 100–102, 115, 140

subsonic parachute, 36, 59

subsystem hardware, 117

Super Lightweight Ablator (SLA), 106

supersonic

  inflatable aerodynamic decelerator, 53

  supersonic parachute, 11, 27, 52, 59, 69, 97, 184, *185*, 212

T

"Tango Delta nominal," xiv

temperature sensors, 174

test bed(s), 39, 116–17, 123, 148, 161–65, 170–73, 176–78, 191, 205–6

  competition among teams to schedule, 123

  computers, 60

  drill, 159

  EDL, 163–65, 170, 176

  electronics, 109

  JPL, 39–40, 60, 109, 116–17, 123, 148, 159, 161–65, 170–73, 176–78, 191, 205–6

  lab, 40

  Mission Systems, 117, 159, 161, 171, 205

  rover, 144, 152, *152*, 159, 161, 177

  shift, 170

  software, 163–64

  spacecraft, 171, 176

vehicle, 177

Vehicle Systems, 116

Theisinger, Pete (MSL project manager), 55–56, 141, 148, 176, 191

thermal protection, 60. *See also* heat shield

three-legged lander, 24, 56. *See also* legged landers

throttled engines (Viking), 28

thrusters, change-of-course, 59

Thurman, Sam (Pathfinder project), 16, 31–32

Tiger Team (group of experts), 55, 67, 94, 134, 165

TLS. *See* Tunable Laser Spectrometer

Tompkins, Vandana "Vandi" (rover planner), 192–94

touchdown detection sensors, 63

touchdown sensor, 57, 62

Trosper, Jennifer (rover operations team), 45–47, 170

Tunable Laser Spectrometer (TLS), 76

turret, 77, *95*, 131–32

two-stage parachutes, 27

U

UHF radio signals, xiv, 177–78

University of Arizona, 119–20, 188

V

Vectran bridle ropes, 186

Vehicle Systems Test Bed, 116

Viking Orbiters and Landers, 8, 24, 39, 87, 89, 185, 214

Viking landing design, 24

W

Watkins, Mike (MSL's mission manager), 153–54

Weiler, Ed (NASA associate administrator), 115, 136

White Sands Proving Grounds (New Mexico), 118

Wiens, Roger (Los Alamos National Laboratory), 79, 81–84, 104, 197, 212

Willis, Jason (EDL systems engineer), 43, 163

Witkowski, Al (Mars parachute builder), 52–53

Y

Yellowknife area and rock analysis, 209